Class Field Theory

Jürgen Neukirch

Class Field Theory

-The Bonn Lectures-
Edited by Alexander Schmidt

Translated from the German
by F. Lemmermeyer and W. Snyder
Language Editor: A. Rosenschon

 Springer

Jürgen Neukirch †

Editor
Alexander Schmidt
Mathematical Institute
Heidelberg University
Germany

ISBN 978-3-642-35436-6   ISBN 978-3-642-35437-3 (eBook)
DOI 10.1007/978-3-642-35437-3
Springer Heidelberg New York Dordrecht London

Library of Congress Control Number: 2013935211

Printed on acid-free paper

Springer is part of Springer Science+Business Media (www.springer.com)

Foreword

In 1969, Jürgen Neukirch's book *Klassenkörpertheorie* was published by *Bibliographisches Institut Mannheim*. The main goal of the book was to grant the reader, who has acquainted himself with the basics of algebraic number theory, a quick and immediate access to class field theory.

Although this book has been out of print for many years, it has remained the favorite introduction to class field theory in Germany. As a student in the 1980s, I myself studied a copy from the library that showed clear signs of extensive use. This motivated the idea to make the text available again, as a printed book as well as a freely accessible file for downloading.

This book should not be confused with Neukirch's book "Class Field Theory" (Springer Grundlehren vol. 280, 1986), which has another focus. The text presented here is essentially identical with the German text based on Neukirch's original Bonn lectures; I only corrected mistakes and updated notation.

I would like to thank Rita Neukirch for her generous support for this new English edition of the work of her late husband. I also thank Rosina Bonn for her excellent typesetting of the original German text into LaTeX and my friends and colleagues Andreas Rosenschon, Bernd Schober and Malte Witte for their invaluable help in improving the English edition.

Heidelberg, October 2012   *Alexander Schmidt*

Dedicated to my teacher

WOLFGANG KRULL

Preface

The present manuscript is an improved edition of a text that first appeared under the same title in *Bonner Mathematische Schriften*, no. 26, and originated from a series of lectures given by the author in 1965/66 in W. KRULL's seminar in Bonn. Since the mathematical literature lacked a uniform presentation of class field theory based on modern cohomological methods, a summarizing exposition of these lectures seemed to be useful. The main goal was to provide the reader, who has acquainted himself with the basics of algebraic number theory, a quick and immediate access to class field theory.

This script consists of three parts, the first of which discusses the cohomology of finite groups. Nowadays, cohomology has conquered large areas of algebraic number theory. Nevertheless, the question whether class field theory can be done without this machinery is a frequent topic of discussion. However, apart from the possibility of formulating the theory in terms of algebras, which is closely related to cohomology, we do not dispose of such a theory at this point, although recent results due to J. LUBIN and J. TATE on the explicit determination of the local norm residue symbol provide some support for this viewpoint. But one must not overlook the fact that cohomology presents – in particular for the learner – a wealth of far reaching advantages. In class field theory, cohomology plays the role of a calculus that allows a clear and logical development of the theory under a unified viewpoint. Its importance, however, is by no means only of formal nature. In fact, local class field theory could originally be developed by defining the norm residue symbol via the Frobenius automorphism for unramified extensions only. It was cohomology that gave a vital impetus to the theory, making also the ramified extensions accessible to class field theoretic methods. This relationship was discovered by H. HASSE and had an immediate impact also on the global theory. Although it was formulated in the language of algebras at first, the cohomological principles behind it did not remain hidden for long. In addition, beyond class field theory, the use of cohomological methods in general field theory has led, via Galois cohomology, to a wealth of far reaching results with a novel allure.

This is another reason why a student may wish to learn the effectiveness of the cohomological calculus in class field theory, it provides him with a concrete example of techniques which are used in many other areas of mathematics.

On the other hand, it cannot be denied that some students interested in class field theory are deterred by cohomology, which at first sight may seem to be some kind of mysterious formal mechanism that is difficult to understand. For this reason we only introduce those notions and results from cohomology that are essential for field theoretic applications, and we have made every endeavor to present the material in a way that is as elementary as possible by avoiding the general notions of homological algebra.

The second part discusses local class field theory. We have put ARTIN's and TATE's theory of class formations at the beginning; it brings out the purely group theoretical formalism of local and global class field theory based on the theorem of TATE. For the sake of formal simplicity we have used the notion of a profinite group; it is, however, not absolutely necessary for an understanding of what follows, since all the essential theorems only refer to finite groups, which are the building blocks of profinite groups. In § 7 we have included the recent results by LUBIN and TATE [34] on the explicit determination of the norm residue symbol, which also will be applied later in the global part of the proof of ARTIN's reciprocity law.

The third part concerns the class field theory of finite algebraic number fields. For the sake of a development that is as straightforward as possible we have decided to omit the theory of function fields over finite constant fields. In order to elaborate to what extend the global theorems can be deduced from their local counterparts, we have strictly separated considerations of a purely local character from those possessing a specific global nature. For a clear presentation it turned out to be appropriate to single out certain cohomology groups that occur when considering different field extensions simultaneously. We have exclusively used CHEVALLEY's notion of idèles for developing the global theory, and yet have tried to emphasize the importance of the classical theory going back to KUMMER. We have obtained a clear structuring of the proof of the reciprocity law by strictly separating the treatment of the idèle group from that of the idèle class group. In the last section we establish the connection between the modern and the classical purely ideal theoretic version of class field theory in the sense of HASSE's *Zahlbericht*.

I would like to thank my honored teacher, Professor W. KRULL, for his active interest and concern in the genesis of this script. K.-O. STÖHR has acquired a special merit to the text; I thank him deeply for his first elaboration of my occasionally rather sketchy accounts on cohomology and local class field theory, as well as for suggesting many essential improvements.

Bonn, July 1969   *Jürgen Neukirch*

Contents

Part I

Cohomology of Finite Groups

J. Neukirch, *Class Field Theory*, DOI 10.1007/978-3-642-35437-3_1,
© Springer-Verlag Berlin Heidelberg 2013

§ 1. *G*-Modules

The cohomology of finite groups deals with a general situation that occurs frequently in different concrete forms. For example, if $L|K$ is a finite Galois extension with Galois group G, then G acts on the multiplicative group $L^\times$ of the extension field L. In the special case of an extension of finite algebraic number fields, G acts on the ideal group J of the extension field L. The theory of group extensions provides us with the following example: If G is an abstract finite group and A is a normal subgroup, then G acts on A via conjugation. In representation theory we study matrix groups G that act on a vector space. The basic notion underlying all these examples is that of a G-module. We will now present some general considerations about G-modules, some of which the reader may already know from the theory of modules over general rings.

In the following, G will always denote a multiplicatively written finite group; its unit element will be denoted by 1.

(1.1) Definition. *A **G-module** A is an (additively written) abelian group A on which the group G acts in such a way that for all $\sigma, \tau \in G$ and all $a, b \in A$ we have*

1) $1a = a$,
2) $\sigma(a + b) = \sigma a + \sigma b$,
3) $(\sigma\tau)a = \sigma(\tau a)$.

Although in applications we mainly deal with multiplicatively written G-modules A, we prefer for formal reasons to write these groups in this section additively.

We can interpret G-modules as modules over rings by introducing the **group ring $\mathbb{Z}[G]$** of G. This ring consists of all formal sums

$$\sum_{\sigma \in G} n_\sigma \sigma$$

with integral coefficients $n_\sigma \in \mathbb{Z}$. In other words, $\mathbb{Z}[G]$ is the free abelian group on the elements of G:

$$\mathbb{Z}[G] = \Big\{ \sum_{\sigma \in G} n_\sigma \sigma \mid n_\sigma \in \mathbb{Z} \Big\}.$$

Since the sums $\sum_{\sigma \in G} n_\sigma \sigma$ can be multiplied, $\mathbb{Z}[G]$ is a ring. Therefore we may interpret a G-module A as a module over the ring $\mathbb{Z}[G]$, where the action of $\mathbb{Z}[G]$ on A is defined as

$$\Big(\sum_{\sigma \in G} n_\sigma \sigma \Big) a = \sum_{\sigma \in G} n_\sigma (\sigma a), \quad a \in A.$$

Of course, $\mathbb{Z}[G]$ considered just as an additive group is a G-module itself; it will play a distinguished role in our considerations.

The group ring $\mathbb{Z}[G]$ contains two distinguished ideals:

$$I_G = \left\{ \sum_{\sigma \in G} n_\sigma \sigma \mid \sum_{\sigma \in G} n_\sigma = 0 \right\} \quad \text{and} \quad \mathbb{Z} \cdot N_G = \left\{ n \cdot \sum_{\sigma \in G} \sigma \mid n \in \mathbb{Z} \right\}.$$

The ideal I_G is called the **augmentation ideal** of $\mathbb{Z}[G]$. It is the kernel of the homomorphism

$$\varepsilon : \mathbb{Z}[G] \longrightarrow \mathbb{Z} \quad \text{with} \quad \varepsilon\left(\sum_{\sigma \in G} n_\sigma \sigma \right) = \sum_{\sigma \in G} n_\sigma,$$

which is called the **augmentation** of $\mathbb{Z}[G]$.

The element $N_G = \sum_{\sigma \in G} \sigma \in \mathbb{Z}[G]$ is called the **norm** (or also **trace**) of $\mathbb{Z}[G]$. For each element $\tau \in G$ we have $\tau N_G = \sum_{\sigma \in G} \tau \sigma = N_G$, which implies that $\mathbb{Z} \cdot N_G$ is an ideal in $\mathbb{Z}[G]$. The map

$$\mu : \mathbb{Z} \longrightarrow \mathbb{Z}[G] \quad \text{with} \quad \mu(n) = n \cdot N_G$$

is called the **coaugmentation** of $\mathbb{Z}[G]$. We set $J_G = \mathbb{Z}[G]/\mathbb{Z} \cdot N_G$, and obtain the **exact sequences**[1] of rings and ring homomorphisms

$$0 \longrightarrow I_G \longrightarrow \mathbb{Z}[G] \xrightarrow{\varepsilon} \mathbb{Z} \longrightarrow 0,$$
$$0 \longrightarrow \mathbb{Z} \xrightarrow{\mu} \mathbb{Z}[G] \longrightarrow J_G \longrightarrow 0.$$

Viewing these rings only as additive groups, we see immediately that they are all free abelian groups, and that I_G and J_G are direct summands of $\mathbb{Z}[G]$:

(1.2) Proposition. *I_G is the free abelian group on the elements $\sigma - 1$, $\sigma \in G$, $\sigma \neq 1$, and J_G is the free abelian group on the elements $\sigma \mod \mathbb{Z} \cdot N_G$, $\sigma \neq 1$. We have direct sum decompositions*

$$\mathbb{Z}[G] = I_G \oplus \mathbb{Z} \cdot 1 \cong I_G \oplus \mathbb{Z},$$
$$\mathbb{Z}[G] = \left(\bigoplus_{\sigma \neq 1} \mathbb{Z}\sigma \right) \oplus \mathbb{Z} \cdot N_G \cong J_G \oplus \mathbb{Z}.$$

Proof. If $\sum_{\sigma \in G} n_\sigma \sigma \in I_G$, then $\sum_{\sigma \in G} n_\sigma = 0$, from which we obtain

$$\sum_{\sigma \in G} n_\sigma \sigma = \sum_{\sigma \in G} n_\sigma (\sigma - 1);$$

if, in addition, $\sum_{\sigma \in G, \sigma \neq 1} n_\sigma (\sigma - 1) = 0$, then $n_\sigma = 0$ for all $\sigma \in G$, $\sigma \neq 1$.

---

[1] A sequence $\cdots \to A \xrightarrow{i} B \xrightarrow{j} C \to \cdots$ of groups, modules or rings and homomorphisms $i, j, \ldots$ is called exact, if the image of each map is equal to the kernel of the subsequent map. We will often have to deal with short exact sequences $0 \to A \xrightarrow{i} B \xrightarrow{j} C \to 0$. Such a sequence encodes the information that the homomorphism $j : B \to C$ is surjective with kernel $iA \cong A$.

Since each element $\sum_{\sigma \in G} n_\sigma \sigma \in \mathbb{Z}[G]$ can be written in the following form

$$\sum_{\sigma \in G} n_\sigma \sigma = \sum_{\sigma \in G} n_\sigma(\sigma - 1) + \left(\sum_{\sigma \in G} n_\sigma\right) \cdot 1,$$

we have the decomposition as an evidently direct sum $\mathbb{Z}[G] = I_G \oplus \mathbb{Z} \cdot 1$.

If, on the other hand, $\sum_{\sigma \in G} n_\sigma \sigma \mod \mathbb{Z} \cdot N_G \in J_G$, then we can write

$$\sum_{\sigma \in G} n_\sigma \sigma = \sum_{\sigma \neq 1}(n_\sigma - n_1)\sigma + n_1 \cdot \sum_{\sigma \in G} \sigma \equiv \sum_{\sigma \neq 1}(n_\sigma - n_1)\sigma \mod \mathbb{Z} \cdot N_G,$$

and $\sum_{\sigma \neq 1} n_\sigma \sigma \in \mathbb{Z} \cdot N_G$ clearly implies $n_\sigma = 0$ for all $\sigma \neq 1$.

Thus J_G is the free abelian group generated by the elements $\sigma \mod \mathbb{Z} \cdot N_G$ with $\sigma \neq 1$. Because of the uniqueness of the representation

$$\sum_{\sigma \in G} n_\sigma \sigma = \sum_{\sigma \neq 1}(n_\sigma - n_1)\sigma + n_1 \cdot N_G,$$

we also obtain the direct sum decomposition $\mathbb{Z}[G] = \left(\bigoplus_{\sigma \neq 1} \mathbb{Z}\sigma\right) \oplus \mathbb{Z} \cdot N_G$.

The ideals I_G and $\mathbb{Z} \cdot N_G$ in $\mathbb{Z}[G]$ are dual to each other in the following sense.

(1.3) Proposition. $I_G = \operatorname{Ann} \mathbb{Z} \cdot N_G$ and $\mathbb{Z} \cdot N_G = \operatorname{Ann} I_G$.

Proof. We have

$$\left(\sum_{\sigma \in G} n_\sigma \sigma\right) \cdot N_G = \sum_{\sigma \in G} n_\sigma(\sigma \cdot N_G) = \sum_{\sigma \in G} n_\sigma N_G = \left(\sum_{\sigma \in G} n_\sigma\right) \cdot N_G = 0$$

if and only if

$$\sum_{\sigma \in G} n_\sigma = 0.$$

Thus $\operatorname{Ann} \mathbb{Z} \cdot N_G = I_G$. On the other hand, we know from (1.2) that I_G is the free abelian group generated by the elements $\sigma - 1$, $\sigma \in G$. Therefore

$$\sum_{\tau \in G} n_\tau \tau \in \operatorname{Ann} I_G \Longleftrightarrow \left(\sum_{\tau \in G} n_\tau \tau\right)(\sigma - 1) = 0 \qquad \text{for all } \sigma \in G$$

$$\Longleftrightarrow \sum_{\tau \in G} n_\tau \tau \sigma = \sum_{\tau \in G} n_\tau \tau \qquad \text{for all } \sigma \in G$$

$$\Longleftrightarrow n_\tau = n_1 \qquad \text{for all } \tau \in G$$

$$\Longleftrightarrow \sum_{\tau \in G} n_\tau \tau = n_1 \cdot N_G \in \mathbb{Z} \cdot N_G,$$

so that $\mathbb{Z} \cdot N_G = \operatorname{Ann} I_G$, as claimed.

After these remarks on group rings we now return to general *G*-modules. For each *G*-module A we have the following four distinguished submodules:

$A^G = \{a \in A \mid \sigma a = a \text{ for all } \sigma \in G\}$, the **fixed group** of A,

$N_G A = \{N_G a = \sum_{\sigma \in G} \sigma a \mid a \in A\}$, the **norm group** of A [2]),

$_{N_G} A = \{a \in A \mid N_G a = 0\}$,

$I_G A = \{\sum_{\sigma \in G} n_\sigma (\sigma a_\sigma - a_\sigma) \mid a_\sigma \in A\}$.

Since I_G is the module generated by the elements $\sigma - 1$, $\sigma \in G$, we obviously have $A^G = \{a \in A \mid I_G a = 0\}$. On the other hand, $I_G A$ is the module generated by the elements $\sigma a - a$, $a \in A$, $\sigma \in G$. Proposition (1.3) provides us with the inclusions

$$N_G A \subseteq A^G \quad \text{and} \quad I_G A \subseteq {_{N_G}} A,$$

and we can form the factor groups

$$A^G / N_G A \quad \text{and} \quad {_{N_G}} A / I_G A.$$

These groups will turn out to be the cohomology groups of the G-module A of dimension 0 and -1 respectively.

If A is a G-module and g is a subgroup of G, it is clear that A is also a g-module. Moreover, if g is a normal subgroup of G, the fixed module A^g is obviously a G/g-module.

In the following we consider the most important functorial properties of G-modules.

Let A and B denote two G-modules. A homomorphism

$$f : A \longrightarrow B$$

is called a **G-homomorphism** if $f(\sigma a) = \sigma f(a)$ for all $\sigma \in G$. We will often interpret a G-module A simply as an abelian group; in this case we will talk about **$\mathbb{Z}$-modules** and **$\mathbb{Z}$-homomorphisms** instead of G-modules and G-homomorphisms.

Given two G-modules A and B we can construct a third G-module, the module

$$\mathrm{Hom}(A, B)$$

of all $\mathbb{Z}$-homomorphisms $f : A \to B$, on which the elements $\sigma \in G$ act as:

$$\sigma(f) = \sigma \circ f \circ \sigma^{-1}, \text{ that is,} \quad \sigma(f)(a) = \sigma f(\sigma^{-1} a), \ a \in A.$$

The group $\mathrm{Hom}_G(A, B)$ of all G-homomorphisms from A to B is a subgroup of $\mathrm{Hom}(A, B)$; clearly it is the fixed module of the G-module $\mathrm{Hom}(A, B)$:

$$\mathrm{Hom}_G(A, B) = \mathrm{Hom}(A, B)^G.$$

---

[2]) For the elements $\sum_{\sigma \in G} \sigma a$ the name **trace** seems more appropriate. However, given that our later applications mainly involve multiplicative G-modules, we have decided use the word **norm** already here.

In addition to $\text{Hom}_G(A, B)$ we have another *G*-module, the **tensor product**

$$A \otimes_{\mathbf{Z}} B.$$

Roughly speaking, the tensor product of A and B consists of all formal sums of products $\sum_i a_i \cdot b_i$, $a_i \in A$, $b_i \in B$. A precise definition is given by:

(1.4) Definition. *Given two abelian groups ($\mathbf{Z}$-modules) A and B, let F be the free abelian group generated by all pairs (a, b), $a \in A$, $b \in B$, and let R denote the subgroup of F generated by the elements of the form*

$$(a + a', b) - (a, b) - (a', b) \quad \text{and} \quad (a, b + b') - (a, b) - (a, b').$$

Then the factor group

$$F/R = A \otimes_{\mathbf{Z}} B$$

*is called the **tensor product** of A and B over $\mathbf{Z}$.*

Since we only consider tensor products over the ring $\mathbf{Z}$, we will write for simplicity $A \otimes B$ instead of $A \otimes_{\mathbf{Z}} B$. We denote by $a \otimes b$ the coset

$$a \otimes b = (a, b) + R \in A \otimes B.$$

By definition, the tensor product $A \otimes B$ consists of all elements of the form

$$\sum_i a_i \otimes b_i, \quad a_i \in A, \; b_i \in B,$$

hence is generated by the elements $a \otimes b$.

In the special case $A = \mathbf{Z}$ we will often regard the tensor product $\mathbf{Z} \otimes B$ and the $\mathbf{Z}$-module B as equal by identifying[3] $n \otimes b$ and $n \cdot b$ for $n \in \mathbf{Z}$, $b \in B$.

If A and B are two abelian groups, we will identify the groups $A \otimes B$ and $B \otimes A$ via the isomorphism

$$f : A \otimes B \longrightarrow B \otimes A \quad \text{with} \quad f(a \otimes b) = b \otimes a.$$

Similarly, if A, B, C are three abelian groups, we will regard the groups $(A \otimes B) \otimes C$ and $A \otimes (B \otimes C)$ as equal, making use of the isomorphism

$$f : (A \otimes B) \otimes C \longrightarrow A \otimes (B \otimes C) \quad \text{with} \quad f((a \otimes b) \otimes c) = a \otimes (b \otimes c).$$

If A and B are two *G*-modules, then $A \otimes B$ becomes a *G*-module by defining

$$\sigma(a \otimes b) = \sigma a \otimes \sigma b, \quad a \in A, \quad b \in B; \quad \sigma \in G.\,[4]$$

---

[3] Starting from this case, we can think of the construction of the general tensor product as a formal change of the domain of coefficients of B from $\mathbf{Z}$ to A. For example, every abelian group B can be extended to a $\mathbf{Q}$-vector space by tensoring B with $\mathbf{Q}$. This transition is the formal extension of the scalars of the $\mathbf{Z}$-module B from $\mathbf{Z}$ to $\mathbf{Q}$, and multiplication of $b \in B$ by a rational number $r \in \mathbf{Q}$ corresponds to forming $r \otimes b \in \mathbf{Q} \otimes B$.

[4] Since the products $a \otimes b$ generate the module $A \otimes B$, we obtain the action of σ on the entire module by linear extension.

In general it is not true that $A^G \otimes B^G$ is the fixed module of $A \otimes B$. In fact $A^G \otimes B^G$ might not even be a submodule of $A \otimes B$. We only have the canonical (but in general neither injective nor surjective) homomorphism

$$A^G \otimes B^G \longrightarrow (A \otimes B)^G.$$

It is easy to verify that the functors Hom_G and $\otimes$ are additive:

(1.5) Proposition. *Let $\{A_\iota \mid \iota \in I\}$ be a family of G-modules, and let X be another G-module. Then we have canonical isomorphisms* [5]

$$X \otimes \left(\bigoplus_\iota A_\iota \right) \cong \bigoplus_\iota (X \otimes A_\iota),$$

$$\mathrm{Hom}_G\left(\bigoplus_\iota A_\iota, X \right) \cong \prod_\iota \mathrm{Hom}_G(A_\iota, X), \ \ \mathrm{Hom}_G\left(X, \prod_\iota A_\iota \right) \cong \prod_\iota \mathrm{Hom}_G(X, A_\iota).$$

Moreover, if X is finitely generated as an abelian group, then

$$X \otimes \left(\prod_\iota A_\iota \right) \cong \prod_\iota (X \otimes A_\iota), \quad \mathrm{Hom}_G\left(X, \bigoplus_\iota A_\iota \right) \cong \bigoplus_\iota \mathrm{Hom}_G(X, A_\iota).$$

Let A and B be G-modules and

$$A \xrightarrow{\ h\ } A'$$

a G-homomorphism. Then h induces a G-homomorphism in the 'opposite' direction

$$\mathrm{Hom}(A, B) \longleftarrow \mathrm{Hom}(A', B)$$

given by composition $f \mapsto f \circ h$ $(f \in \mathrm{Hom}(A', B))$, and a G-homomorphism

$$A \otimes B \longrightarrow A' \otimes B$$

defined by $a \otimes b \mapsto h(a) \otimes b$. On the other hand, given a G-homomorphism

$$B \xrightarrow{\ g\ } B',$$

there are analogously defined G-homomorphisms

$$\mathrm{Hom}(A, B) \longrightarrow \mathrm{Hom}(A, B')$$

and

$$A \otimes B \longrightarrow A \otimes B'.$$

Because of these properties, Hom is called a contravariant functor in the first and a covariant functor in the second argument, and the tensor product $\otimes$ is called a covariant functor in both arguments.

---

[5] The symbol $\bigoplus$ stands for the **direct sum**, i.e., the group of all families $(\ldots, a_\iota, \ldots)$, where only finitely many components a_ι are different from zero. In contrast, $\prod$ denotes the **direct product**, i.e., the group of all families $(\ldots, a_\iota, \ldots)$.

If we have two G-homomorphisms

$$A' \xrightarrow{h} A \quad \text{and} \quad B \xrightarrow{g} B',$$

we obtain a G-homomorphism

$$(h, g) : \text{Hom}(A, B) \longrightarrow \text{Hom}(A', B')$$

by defining $f \mapsto g \circ f \circ h$ ($f \in \text{Hom}(A, B)$), and for two G-homomorphisms

$$A \xrightarrow{h} A' \quad \text{and} \quad B \xrightarrow{g} B'$$

we get a G-homomorphism

$$h \otimes g : A \otimes B \longrightarrow A' \otimes B',$$

by setting $h \otimes g \, (a \otimes b) = h(a) \otimes g(b)$.

In what follows, G-free G-modules will play an important role. A G-module A is called **G-free** or **$\mathbb{Z}[G]$-free**, if it is the direct sum of G-modules which are isomorphic to $\mathbb{Z}[G]$. A first basic property of G-free modules is:

(1.6) Proposition. *Let X be a G-free G-module and*

$$0 \longrightarrow A \xrightarrow{h} B \xrightarrow{g} C \longrightarrow 0$$

an exact sequence of G-modules A, B, C and G-homomorphisms h, g. Then the induced sequence

$$0 \longrightarrow \text{Hom}_G(X, A) \longrightarrow \text{Hom}_G(X, B) \longrightarrow \text{Hom}_G(X, C) \longrightarrow 0$$

is also exact.

Write $X = \bigoplus_\iota \Gamma_\iota$ with $\Gamma_\iota \cong \mathbb{Z}[G]$. From (1.5) we have the decomposition

$$\text{Hom}_G(X, A) = \prod_\iota \text{Hom}_G(\Gamma_\iota, A).$$

If we set $A_\iota = \text{Hom}_G(\Gamma_\iota, A) \cong \text{Hom}_G(\mathbb{Z}[G], A) \cong A$ (where the last isomorphism is given by the map $f \in \text{Hom}_G(\mathbb{Z}[G], A) \mapsto f(1) \in A$), and define B_ι and C_ι analogously, we obtain the exact sequence

$$0 \longrightarrow A_\iota \longrightarrow B_\iota \longrightarrow C_\iota \longrightarrow 0,$$

which implies the proposition.

Remark. Proposition (1.6) is valid more generally for so-called **projective** G-modules X, i.e., G-modules which have the property that every diagram

$$
\begin{array}{ccc}
 & X & \\
f \downarrow & \searrow f' & \\
B & \xrightarrow{g} C \longrightarrow 0 &
\end{array}
$$

with G-modules B, C and G-homomorphisms g, f', g surjective, can be extended to a commutative diagram by a G-homomorphism $f : X \to B$.

For arbitrary G-modules X it is easy to verify that if one omits the last map $\to 0$ in the induced Hom_G-sequence, the remaining sequence is still exact.

A G-free G-module is, of course, also $\mathbb{Z}$-free, and therefore a free abelian group, since $\mathbb{Z}[G]$ is the free abelian group generated by the elements of G. For most questions concerning the exactness of sequences it suffices to consider only $\mathbb{Z}$-modules and $\mathbb{Z}$-homomorphisms. In later applications we will need the following three lemmas.

(1.7) Lemma. *If* $\cdots \longleftarrow X_{q-1} \overset{d_q}{\longleftarrow} X_q \overset{d_{q+1}}{\longleftarrow} X_{q+1} \longleftarrow \cdots$ *is an exact sequence of $\mathbb{Z}$-free modules and D is an arbitrary $\mathbb{Z}$-module, then the sequence*

$$\cdots \longrightarrow \mathrm{Hom}(X_{q-1}, D) \longrightarrow \mathrm{Hom}(X_q, D) \longrightarrow \mathrm{Hom}(X_{q+1}, D) \longrightarrow \cdots$$

is also exact.

Proof. Let $C_q = \ker d_q = \operatorname{im} d_{q+1}$. Since C_{q-1} is free as a subgroup of X_{q-1}, the exact sequence $0 \leftarrow C_{q-1} \leftarrow X_q \leftarrow C_q \leftarrow 0$ is split, i.e., there is a homomorphism $\varepsilon : C_{q-1} \to X_q$ with $d_q \circ \varepsilon = \mathrm{id}$, and C_q is a direct summand of $X_q : X_q = C_q \oplus X_q'$ for all q. Thus if f is in the kernel of $\mathrm{Hom}(X_q, D) \to \mathrm{Hom}(X_{q+1}, D)$, then f vanishes on C_q, and therefore induces a homomorphism $g' : C_{q-1} \to D$ with $f = g' \circ d_q$. Since C_{q-1} is a direct summand of X_{q-1}, we can extend g' to a homomorphism $g \in \mathrm{Hom}(X_{q-1}, D)$, and f is the image of g under the homomorphism $\mathrm{Hom}(X_{q-1}, D) \to \mathrm{Hom}(X_q, D)$. On the other hand, if $f \in \mathrm{Hom}(X_q, D)$ is in the image of $\mathrm{Hom}(X_{q-1}, D) \to \mathrm{Hom}(X_q, D)$, $f = f' \circ d_q$ with $f' \in \mathrm{Hom}(X_{q-1}, D)$, then $f \circ d_{q+1} = f' \circ d_q \circ d_{q+1} = 0$, i.e., f lies in the kernel of $\mathrm{Hom}(X_q, D) \to \mathrm{Hom}(X_{q+1}, D)$.

(1.8) Lemma. *If $0 \to X \to Y \to Z \to 0$ is an exact sequence of free $\mathbb{Z}$-modules and A is an arbitrary $\mathbb{Z}$-module, then the sequence*

$$0 \longrightarrow X \otimes A \longrightarrow Y \otimes A \longrightarrow Z \otimes A \longrightarrow 0$$

is also exact.

Proof. The exactness of the sequence $X \otimes A \to Y \otimes A \to Z \otimes A \to 0$ is completely trivial and holds without assuming that the modules are free. Hence we only need to show that the map $X \otimes A \to Y \otimes A$ is injective. Since Z is free, there is a homomorphism $Z \to Y$ whose composite with the given map $Y \to Z$ is the identity map on Z. This implies that the image X' of X in Y is a direct summand, i.e., $Y = X' \oplus X''$. Thus we find $Y \otimes A = (X' \otimes A) \oplus (X'' \otimes A)$, which implies the claimed injectivity.

(1.9) Lemma. *If $0 \to A \to B \to C \to 0$ is an exact sequence of $\mathbb{Z}$-modules and X a $\mathbb{Z}$-free $\mathbb{Z}$-module, then*

$$0 \longrightarrow X \otimes A \longrightarrow X \otimes B \longrightarrow X \otimes C \longrightarrow 0$$

is also an exact sequence.

Proof. Let $X = \bigoplus_\iota Z_\iota$ with $Z_\iota \cong \mathbb{Z}$. Since the functor $\otimes$ is additive, we have the canonical isomorphism

$$X \otimes A \cong \bigoplus_\iota (Z_\iota \otimes A) \cong \bigoplus_\iota A_\iota \quad \text{with} \quad A_\iota = Z_\iota \otimes A \cong A,$$

and similarly for B and C. Hence the exactness of the sequence

$$0 \longrightarrow A_\iota \longrightarrow B_\iota \longrightarrow C_\iota \longrightarrow 0$$

immediately implies the exactness of

$$0 \longrightarrow X \otimes A \longrightarrow X \otimes B \longrightarrow X \otimes C \longrightarrow 0.$$

§ 2. The Definition of Cohomology Groups

It is a characteristic of cohomology theory that in order to give even simple definitions and theorems, one must introduce an extensive formalism of homomorphisms, functors and sequences. At first this might give the impression that we are dealing with a particularly difficult and deep mathematical discipline. However, once the reader has become more familiar with these methods, he will realize that these considerations are particularly simple, and they may even appear somewhat anemic. Nevertheless, their frequent repeated use leads to concepts and theorems which could hardly be developed using a more elementary approach. In order to introduce our cohomology groups, we start with such formal considerations, although the definition of these groups can be given in a direct and elementary way, as well[6].

Let G be a finite group. By a **complete free resolution** of the group G, or also of the G-module $\mathbb{Z}$ [7], we mean a complex

$$\cdots \xleftarrow{d_{-2}} X_{-2} \xleftarrow{d_{-1}} X_{-1} \xleftarrow{d_0} X_0 \xleftarrow{d_1} X_1 \xleftarrow{d_2} X_2 \xleftarrow{d_3} \cdots$$

with the following properties:

----

[6] Cf. [16], 15.7, p. 236.

[7] We always consider $\mathbb{Z}$ as a G-module by letting the group G act on $\mathbb{Z}$ trivially (i.e., as the identity).

(1) The X_q are free G-modules,
(2) ε, μ, d_q are G-homomorphisms,
(3) $d_0 = \mu \circ \varepsilon$,
(4) at every term we have exactness.

Thus a complete free resolution consists in fact of two exact sequences

$$0 \longleftarrow \mathbb{Z} \xleftarrow{\ \varepsilon\ } X_0 \xleftarrow{\ d_1\ } X_1 \xleftarrow{\ d_2\ } \cdots$$

and

$$0 \longrightarrow \mathbb{Z} \xrightarrow{\ \mu\ } X_{-1} \xrightarrow{\ d_{-1}\ } X_{-2} \xrightarrow{\ d_{-2}\ } \cdots$$

of free G-modules which are spliced together. Originally, cohomology groups were derived from the first, and homology groups from the second sequence. But putting these two sequences together is a crucial step which leads to a unified and, concerning the functorial properties, harmonious fusion of homology and cohomology.

When defining the cohomology groups of a G-module we could start with an arbitrary complete resolution, even one in which the X_q need only be projective G-modules. But then we would need to show the definition is independent from the chosen resolution. In order to avoid this effort, we start with a particular complete resolution, the so-called **standard resolution**, which has its origins in algebraic topology and arises in the following way.

For $q \geq 1$ we consider all q-tuples $(\sigma_1, \ldots, \sigma_q)$, where the σ_i run through the group G; we call such a q-tuple a q-**cell** (with "vertices" $\sigma_1, \ldots, \sigma_q$). We use these q-cells as free generators of our G-module, i.e., we set

$$X_q = X_{-q-1} = \bigoplus \mathbb{Z}[G](\sigma_1, \ldots, \sigma_q).$$

For $q = 0$ we put

$$X_0 = X_{-1} = \mathbb{Z}[G],$$

where we choose the identity element $1 \in \mathbb{Z}[G]$ as the generating "null cell". In particular, the modules

$$\ldots, X_{-2}, X_{-1}, X_0, X_1, X_2, \ldots$$

are free G-modules.

Define the G-homomorphisms $\varepsilon : X_0 \to \mathbb{Z}$ and $\mu : \mathbb{Z} \to X_{-1}$ by (cf. §1, p. 4)

$$\varepsilon(\textstyle\sum_{\sigma \in G} n_\sigma \sigma) = \sum_{\sigma \in G} n_\sigma \qquad \text{(augmentation)}$$
$$\mu(n) = \quad n \cdot N_G \qquad \text{(coaugmentation)}$$

In order to determine the remaining G-homomorphisms d_q it suffices, of course, to give their values on the free generators $(\sigma_1, \ldots, \sigma_q)$. Here we set

$$d_0 1 = N_G \qquad\qquad\qquad\qquad\qquad \text{for } q = 0,$$

$$d_1(\sigma) = \sigma - 1 \qquad\qquad\qquad\qquad\qquad \text{for } q = 1,$$

$$d_q(\sigma_1,\ldots,\sigma_q) = \sigma_1(\sigma_2,\ldots,\sigma_q)$$
$$+ \sum_{i=1}^{q-1}(-1)^i(\sigma_1,\ldots,\sigma_{i-1},\sigma_i\sigma_{i+1},\sigma_{i+2},\ldots,\sigma_q)$$
$$+(-1)^q(\sigma_1,\ldots,\sigma_{q-1}) \qquad\qquad \text{for } q > 1,$$

$$d_{-1}1 = \sum_{\sigma \in G}[\sigma^{-1}(\sigma) - (\sigma)] \qquad\qquad\qquad \text{for } q = -1,$$

$$d_{-q-1}(\sigma_1,\ldots,\sigma_q) = \sum_{\sigma \in G}\sigma^{-1}(\sigma,\sigma_1,\ldots,\sigma_q)$$
$$+ \sum_{\sigma \in G}\sum_{i=1}^{q}(-1)^i(\sigma_1,\ldots,\sigma_{i-1},\sigma_i\sigma,\sigma^{-1},\sigma_{i+1},\ldots,\sigma_q)$$
$$+ \sum_{\sigma \in G}(-1)^{q+1}(\sigma_1,\ldots,\sigma_q,\sigma) \qquad \text{for } -q-1 < -1.$$

From these definitions we obtain a complex

$$\cdots \xleftarrow{d_{-2}} X_{-2} \xleftarrow{d_{-1}} X_{-1} \xleftarrow{d_0} X_0 \xleftarrow{d_1} X_1 \xleftarrow{d_2} X_2 \xleftarrow{d_3} \cdots$$

which we call the **standard complex** of the group G. We will see that this complex is a complete free resolution of the group G. The conditions (1)–(3) are trivially satisfied: By construction the X_q are free G-modules, the ε, μ, d_q are G-homomorphisms; and because $\mu \circ \varepsilon(1) = \mu(1) = N_G = d_0 1$, we have $d_0 = \mu \circ \varepsilon$. Hence it only remains to show that we have exactness at every term. To prove this fact, we use computations that are based on similar considerations in algebraic topology. We first show that the sequence

$$(*) \qquad\qquad 0 \longleftarrow \mathbb{Z} \xleftarrow{\varepsilon} X_0 \xleftarrow{d_1} X_1 \xleftarrow{d_2} X_2 \xleftarrow{d_3} \cdots$$

is exact. To do this, we define the following $\mathbb{Z}$-homomorphisms:

$$E : \mathbb{Z} \longrightarrow X_0 \qquad \text{with } E(1) = 1,$$
$$D_0 : X_0 \longrightarrow X_1 \qquad \text{with } D_0(\sigma) = (\sigma),$$
$$D_q : X_q \longrightarrow X_{q+1} \quad \text{with } D_q(\sigma_0(\sigma_1,\ldots,\sigma_q)) = (\sigma_0,\ldots,\sigma_q) \text{ for } q \geq 1.$$

An elementary calculation shows that

$$E \circ \varepsilon + d_1 \circ D_0 = \text{id} \quad \text{and} \quad D_{q-1} \circ d_q + d_{q+1} \circ D_q = \text{id}.$$

These formulas imply that given $x \in \ker \varepsilon$ (resp. $x \in \ker d_q$), we have $x = d_1 D_0 x \in \text{im } d_1$ (resp. $x = d_{q+1} D_q x \in \text{im } d_{q+1}$), which proves the inclusions

$$\ker \varepsilon \subseteq \text{im } d_1 \quad \text{and} \quad \ker d_q \subseteq \text{im } d_{q+1} \qquad \text{for } q \geq 1.$$

On the other hand, it is easily checked that $\varepsilon \circ d_1 = 0$, so that $\ker \varepsilon \supseteq \text{im } d_1$. We now prove by induction that $d_q \circ d_{q+1} = 0$. For this we assume $d_{q-1} \circ d_q = 0$. (For $q = 1$ we replace d_0 by ε and D_{-1} by E.) Then, on one hand, we have

$$d_q = (D_{q-2} \circ d_{q-1} + d_q \circ D_{q-1}) \circ d_q = d_q \circ D_{q-1} \circ d_q,$$

and on the other hand,

$$d_q = d_q \circ (D_{q-1} \circ d_q + d_{q+1} \circ D_q) = d_q \circ D_{q-1} \circ d_q + d_q \circ d_{q+1} \circ D_q.$$

Subtracting these equations we obtain

$$d_q \circ d_{q+1} \circ D_q = 0.$$

But since every cell in X_{q+1} lies in the image of D_q, we have

$$d_q \circ d_{q+1} = 0,$$

and therefore $\ker d_q \supseteq \operatorname{im} d_{q+1}$ for $q \geq 1$. Thus the sequence $(*)$ is exact.

The second sequence

$$(**) \qquad 0 \longrightarrow \mathbb{Z} \xrightarrow{\mu} X_{-1} \xrightarrow{d_{-1}} X_{-2} \xrightarrow{d_{-2}} X_{-3} \xrightarrow{d_{-3}} \cdots$$

arises from $(*)$ by dualizing. In fact, from $(*)$ we first obtain the sequence

$$(***) \qquad 0 \longrightarrow \operatorname{Hom}(\mathbb{Z}, \mathbb{Z}) \longrightarrow \operatorname{Hom}(X_0, \mathbb{Z}) \longrightarrow \operatorname{Hom}(X_1, \mathbb{Z}) \longrightarrow \cdots,$$

which is exact by (1.7).

Let $\{x_i\}$ be the system of $\mathbb{Z}[G]$-free generators of X_q consisting of all q-cells. Then the "dual basis" $\{x_i^*\}$ of $\{x_i\}$ defined by

$$x_i^*(\sigma x_k) = \begin{cases} 1 & \text{for } \sigma = 1 \text{ and } i = k \\ 0 & \text{otherwise.} \end{cases}$$

is a $\mathbb{Z}[G]$-free system of generators of $\operatorname{Hom}(X_q, \mathbb{Z})$. Thus the G-modules $\operatorname{Hom}(X_q, \mathbb{Z})$ and X_q are canonically isomorphic. If we identify x_i with x_i^*, we can write

$$X_{-q-1} = \operatorname{Hom}(X_q, \mathbb{Z}) \quad (q \geq 0) \quad \text{and} \quad \mathbb{Z} = \operatorname{Hom}(\mathbb{Z}, \mathbb{Z}).$$

An elementary calculation now shows that under these identifications the sequence $(***)$ is transformed into the sequence $(**)$, hence $(**)$ is exact.

Finally, since μ is injective, ε is surjective, and $d_0 = \mu \circ \varepsilon$, we have $\ker d_0 = \ker \varepsilon$ and $\operatorname{im} d_0 = \operatorname{im} \mu$; therefore

$$\ker d_0 = \operatorname{im} d_1 \quad \text{and} \quad \operatorname{im} d_0 = \ker d_{-1}.$$

This completes the proof that the standard complex is exact at all terms.

Now we define our cohomology groups using the standard complex. If A is a G-module, we set

$$A_q = \operatorname{Hom}_G(X_q, A).$$

We call the elements of A_q, i.e., the G-homomorphisms $x : X_q \to A$, the **q-cochains** of A. From the exact sequence

$$\cdots \xleftarrow{d_{-2}} X_{-2} \xleftarrow{d_{-1}} X_{-1} \xleftarrow{d_0} X_0 \xleftarrow{d_1} X_1 \xleftarrow{d_2} X_2 \xleftarrow{d_3} \cdots$$

we obtain the sequence

$$\cdots \xrightarrow{\partial_{-2}} A_{-2} \xrightarrow{\partial_{-1}} A_{-1} \xrightarrow{\partial_0} A_0 \xrightarrow{\partial_1} A_1 \xrightarrow{\partial_2} A_2 \xrightarrow{\partial_3} \cdots ,$$

in which, because $d_q \circ d_{q+1} = 0$, we evidently have $\partial_{q+1} \circ \partial_q = 0$; therefore

$$\operatorname{im} \partial_q \subseteq \ker \partial_{q+1}.$$

Contrary to the first sequence, the second sequence is not exact in general, and the cohomology groups "measure" its deviation from being exact. We set

$$Z_q = \ker \partial_{q+1}, \quad R_q = \operatorname{im} \partial_q ,$$

and call the elements in Z_q (resp. R_q) the **q-cocyles** (resp. **q-coboundaries**).

We now define the cohomology groups of G with coefficients in A as follows:

(2.1) Definition. *The factor group*

$$H^q(G, A) = Z_q / R_q$$

is called the **cohomology group of dimension q** *($q \in \mathbb{Z}$) of the G-module A; we also say $H^q(G, A)$ is the q-th cohomology group with coefficients in A.*

We remark that the cohomology groups $H^{-q-1}(G, A)$ are the usual **homology groups** denoted by $H_q(G, A)$ ($q \geq 1$). In algebraic topology, the cohomology groups (with coefficients in $\mathbb{Z}$) were originally introduced as the character groups of the homology groups. This origin has left traces in the fact that the left side of the standard complex is obtained from the right side by duality. We point out again that splicing the two sides into a complete resolution, which allows interpreting homology groups as cohomology groups of negative dimension, is a crucial step which yields more than just a formal unification[8].

We now come to the problem of analyzing the concrete meaning of cohomology groups. The cochain group

$$A_q = A_{-q-1} = \operatorname{Hom}_G(X_q, A), \quad q \geq 1,$$

consists of all G-homomorphisms $x : X_q \to A$. Since the X_q have the q-cells $(\sigma_1, \ldots, \sigma_q)$ as free generators, a G-homomorphism $x : X_q \to A$ is uniquely determined by its values on the q-tuples $(\sigma_1, \ldots, \sigma_q)$. Thus we can view every cochain as a function with arguments in G and values in A, hence as a map

$$x : \underbrace{G \times \cdots \times G}_{q\text{-times}} \longrightarrow A.$$

By taking this view, we can identify

$$A_q = A_{-q-1} = \{x : \underbrace{G \times \cdots \times G}_{q\text{-times}} \longrightarrow A\}, \quad q \geq 1,$$

---

[8] This fusion of homology and cohomology is due to J. TATE.

and obviously
$$A_0 = A_{-1} = \mathrm{Hom}_G(\mathbb{Z}[G], A) = A.$$

From the definition of the homomorphisms d_q of the standard complex, we obtain for the maps ∂_q in the sequence
$$\cdots \xrightarrow{\partial_{-2}} A_{-2} \xrightarrow{\partial_{-1}} A_{-1} \xrightarrow{\partial_0} A_0 \xrightarrow{\partial_1} A_1 \xrightarrow{\partial_2} A_2 \xrightarrow{\partial_3} \cdots$$

the following formulas

$$\partial_0 x = N_G x \qquad\qquad\qquad\qquad\qquad \text{for } x \in A_{-1} = A,$$

$$(\partial_1 x)(\sigma) = \sigma x - x \qquad\qquad\qquad\qquad \text{for } x \in A_0 = A,$$

$$(\partial_q x)(\sigma_1, \ldots, \sigma_q) = \sigma_1 x(\sigma_2, \ldots, \sigma_q)$$
$$+ \sum_{i=1}^{q-1}(-1)^i x(\sigma_1, \ldots, \sigma_i \sigma_{i+1}, \ldots, \sigma_q)$$
$$+ (-1)^q x(\sigma_1, \ldots, \sigma_{q-1}) \qquad \text{for } x \in A_{q-1},\ q \geq 1,$$

$$\partial_{-1} x = \sum_{\sigma \in G}(\sigma^{-1} x(\sigma) - x(\sigma)) \qquad\qquad \text{for } x \in A_{-2},$$

$$(\partial_{-q-1} x)(\sigma_1, \ldots, \sigma_q) = \sum_{\sigma \in G}\big[\sigma^{-1} x(\sigma, \sigma_1, \ldots, \sigma_q)$$
$$+ \sum_{i=1}^q (-1)^i(\sigma_1, \ldots, \sigma_{i-1}, \sigma_i \sigma, \sigma^{-1}, \sigma_{i+1}, \ldots, \sigma_q)$$
$$+ (-1)^{q+1} x(\sigma_1, \ldots, \sigma_q, \sigma)\big] \quad \text{for } x \in A_{-q-2}, q \geq 0.$$

The q-cocycles are therefore the maps
$$x : G \times \cdots \times G \longrightarrow A$$

with $\partial_{q+1} x = 0$, and the q-coboundaries among them are those maps for which there is a $y \in A_{q-1}$ such that $x = \partial_q y$.

We remark that in algebraic applications only the cohomology groups of low dimension appear. The reason for this is that only for these we have a concrete algebraic interpretation. The cohomological calculus would doubtlessly acquire considerably more significance if we had a tangible interpretation for the cohomology groups of higher dimensions as well. For small dimensions, the cohomology groups are given as follows:

The Group $H^{-1}(G, A)$. We have

$$Z_{-1} = \ker \partial_0 = {}_{N_G}A \qquad ((-1)\text{-cocycles}),$$
$$R_{-1} = \mathrm{im}\, \partial_{-1} = I_G A \qquad ((-1)\text{-coboundaries}).$$

We thus obtain (cf. §1, p. 6)

$$H^{-1}(G, A) = {}_{N_G}A / I_G A.$$

The Group $H^0(G, A)$. We have

$$Z_0 = \ker \partial_1 = A^G \qquad \text{(0-cocycles)},$$
$$R_0 = \operatorname{im} \partial_0 = N_G A \qquad \text{(0-coboundaries)}.$$

We obtain

$$H^0(G, A) = A^G / N_G A,$$

the **norm residue group** of the G-module A; this group is the main object of interest in class field theory.

The Group $H^1(G, A)$. The **1-cocycles** are the functions $x : G \to A$ with $\partial_2 x = 0$, thus satisfying the property

$$x(\sigma\tau) = \sigma x(\tau) + x(\sigma) \quad \text{for } \sigma, \tau \in G.$$

Because this relation is similar to the one of being a homomorphism, the 1-cocycles are often also called **crossed homomorphisms**.

The **1-coboundaries** are obviously the functions

$$x(\sigma) = \sigma a - a, \quad \sigma \in G,$$

with fixed $a \in A = A_0$ (i.e., $x = \partial_1 a$).

If the group G acts trivially (i.e., as the identity) on A, then obviously $Z_1 = \operatorname{Hom}(G, A)$ and $R_1 = 0$; therefore

$$H^1(G, A) = \operatorname{Hom}(G, A).$$

In particular, in case $A = \mathbb{Q}/\mathbb{Z}$, we obtain the **character group** of G:

$$H^1(G, \mathbb{Q}/\mathbb{Z}) = \operatorname{Hom}(G, \mathbb{Q}/\mathbb{Z}) = \chi(G).$$

When studying G-modules and their properties, one is immediately led to the cohomology group $H^1(G, A)$ because of the following considerations:

If we start with a short exact sequence of G-modules A, B, C,

$$0 \longrightarrow A \xrightarrow{i} B \xrightarrow{j} C \longrightarrow 0$$

and pass to the sequence of their fixed modules A^G, B^G, C^G, the sequence

$$0 \longrightarrow A^G \xrightarrow{i} B^G \xrightarrow{j} C^G$$

always remains exact, but the homomorphism j is in general no longer surjective. The question why exactness fails at the last term leads to a canonical homomorphism $C^G \xrightarrow{\delta} H^1(G, A)$.

In fact, let $c \in C^G$. Since the homomorphism $B \xrightarrow{j} C$ is surjective, there is an element $b \in B$ with $jb = c$, but it is not certain that this element b can be chosen from B^G, i.e., in such a way that $\sigma b - b = 0$ for all $\sigma \in G$. Since

$$j(\sigma b - b) = \sigma(jb) - jb = \sigma c - c = 0,$$

we can only say that $\sigma b - b$ lies in the kernel of $B \xrightarrow{j} C$, and thus in the image of $A \xrightarrow{i} B$, so that

$$ia_\sigma = \sigma b - b, \quad a_\sigma \in A.$$

It is easy to verify that a_σ defines a 1-cocycle with coefficients in A. The only freedom in the association $c \mapsto a_\sigma$ is the choice of b with $jb = c$. If we choose a different element b', we obtain a 1-cocycle a'_σ that differs from a_σ only by a 1-coboundary. Therefore every $c \in C^G$ defines a unique cohomology class $\bar{a}_\sigma \in H^1(G, A)$, and it is easy to see that c is in the image of $B^G \to C^G$ if and only if $\bar{a}_\sigma = 0$. In other words, we have a canonical homomorphism $C^G \xrightarrow{\delta} H^1(G, A)$ with the property that the following sequence is exact

$$0 \longrightarrow A^G \xrightarrow{i} B^G \xrightarrow{j} C^G \xrightarrow{\delta} H^1(G, A).$$

In the next section we will encounter these ideas again in a more general setting.

The Group $H^2(G, A)$. We have as **2-cocycles** all functions $x : G \times G \to A$ which satisfy the equation $\partial_3 x = 0$, thus the equation

$$x(\sigma\tau, \rho) + x(\sigma, \tau) = \sigma x(\tau, \rho) + x(\sigma, \tau\rho), \quad \sigma, \tau, \rho \in G.$$

Among these we find that the **2-coboundaries** are the functions such that

$$x(\sigma, \tau) = \sigma y(\tau) - y(\sigma\tau) + y(\sigma)$$

for an arbitrary 1-cochain $y : G \to A$.

Long before the development of cohomology theory the 2-cocycles were known in the theory of groups and algebras as so-called **factor systems**, and it is fair to say that they historically represent the beginning of cohomological considerations in algebra. We want to explain briefly how factor systems come up in the theory of **group extensions**. This is the following type of problem.

Suppose that we are given an abelian group A, written multiplicatively, and an arbitrary group G. We want to find all group extensions $\widehat{G}$ of A (i.e., all groups $\widehat{G}$ with a subgroup isomorphic to A) such that A is a normal subgroup in $\widehat{G}$ and $\widehat{G}/A \cong G$. The question now is which data (except A and G) determine the possible solutions of this problem.

Assume first we have a solution $\widehat{G}$, so that $A \lhd \widehat{G}$ and $\widehat{G}/A \cong G$. If we choose a system of representatives for the factor group $\widehat{G}/A \cong G$, i.e., for each $\sigma \in G$ we choose a preimage $u_\sigma \in \widehat{G}$, then every element in $\widehat{G}$ can be written uniquely in the form

(1) $a \cdot u_\sigma, \quad a \in A, \ \sigma \in G.$

In order to obtain the complete group table for the group $\widehat{G}$, it obviously suffices to know how the products $u_\sigma \cdot a$ ($\sigma \in G$, $a \in A$) and $u_\sigma \cdot u_\tau$ ($\sigma, \tau \in G$) are represented in the form (1).

Now since A is normal in $\widehat{G}$, $u_\sigma \cdot a$ lies in the same right coset as u_σ, i.e.,

$$(2) \qquad u_\sigma \cdot a = a^\sigma \cdot u_\sigma$$

for some $a^\sigma \in A$. This gives the abelian group A a natural structure as a G-module, because the elements $\sigma \in G$ act on A (independently of the choice of u_σ) via the rule $a \mapsto a^\sigma = u_\sigma \cdot a \cdot u_\sigma^{-1}$.

The product $u_\sigma \cdot u_\tau$ lies in the same coset as $u_{\sigma\tau}$, i.e.,

$$(3) \qquad u_\sigma \cdot u_\tau = x(\sigma, \tau) \cdot u_{\sigma\tau} \quad \text{with } x(\sigma, \tau) \in A.$$

In this equation, the factor system $x(\sigma, \tau)$ appears, which is easily seen to be a 2-cocycle of the G-module A. In fact, since multiplication in $\widehat{G}$ is associative,

$$(u_\sigma \cdot u_\tau) \cdot u_\rho = u_\sigma \cdot (u_\tau \cdot u_\rho),$$

from which we get the equation

$$(u_\sigma \cdot u_\tau) \cdot u_\rho = x(\sigma, \tau) \cdot u_{\sigma\tau} \cdot u_\rho = x(\sigma, \tau) \cdot x(\sigma\tau, \rho) \cdot u_{\sigma\tau\rho} =$$
$$u_\sigma \cdot (u_\tau \cdot u_\rho) = u_\sigma \cdot x(\tau\rho) \cdot u_{\tau,\rho} = x^\sigma(\tau, \rho) \cdot u_\sigma \cdot u_{\tau\rho}$$
$$= x^\sigma(\tau, \rho) \cdot x(\sigma, \tau\rho) \cdot u_{\sigma\tau\rho}.$$

It follows that

$$x(\sigma, \tau) \cdot x(\sigma\tau, \rho) = x^\sigma(\tau, \rho) \cdot x(\sigma, \tau\rho),$$

which is precisely the cocycle property.

The above analysis makes use of the choice of a system of representatives u_σ. Given a different system of representatives u'_σ, we get from the equation

$$u'_\sigma \cdot u'_\tau = x'(\sigma, \tau) \cdot u'_{\sigma\tau}$$

another factor system $x'(\sigma, \tau)$. However, this system differs from $x(\sigma, \tau)$ only by a 2-coboundary, namely by the 2-coboundary $\partial_2(u'_\sigma \cdot u_\sigma^{-1})$, as is easily verified. Since the group $\widehat{G}$ is completely determined by the relations (2) and (3) (and by the relations in the groups A and G), this allows us to conclude:

The solution $\widehat{G}$ of the group extension problem is uniquely determined by the action of the group G on A, and a class of equivalent factor systems $x(\sigma, \tau)$, i.e., a cohomology class in $H^2(G, A)$.

Conversely, if we have made the group A into a G-module in any possible way[9], and we are given a class $c \in H^2(G, A)$, we always obtain a solution of the extension problem: For the $\sigma \in G$ choose generators u_σ and define $\widehat{G}$ as the group generated by the elements of A and the u_σ, subject to the relations

$$a^\sigma = u_\sigma \cdot a \cdot u_\sigma^{-1} \quad \text{and} \quad u_\sigma \cdot u_\tau = x(\sigma, \tau) \cdot u_{\sigma\tau} \quad (x(\sigma, \tau) \text{ a 2-cocycle in } c).$$

Again this can be verified very easily.

---

[9] More precisely: Recall that for a G-module A every element $\sigma \in G$ provides an automorphism of the group A. Thus a G-module A is precisely a pair of groups A and G, together with a homomorphism $h : G \to \mathrm{Aut}(A)$; the action of $\sigma \in G$ on A is given by $\sigma a = h(\sigma)a$. Thus making the group A into a G-module simply means choosing a homomorphism from G to the automorphism group $\mathrm{Aut}(A)$.

Together with the groups $H^q(G, A)$ of dimension $q = -1, 0, 1, 2$, the cohomology group $H^{-2}(G, \mathbb{Z})$ with coefficients in $\mathbb{Z}$ plays a special role. We will show later that it is canonically isomorphic to the **abelianization** $G^{\mathrm{ab}} = G/G'$ (where G' is the commutator subgroup) of G, a fact which is heavily used in class field theory. The main theorem of class field theory concerns an isomorphism between the abelianization G^{ab} and the norm residue group $A^G/N_G A$ of a particular G-module A. It can be formulated in purely cohomological terms as $H^{-2}(G, \mathbb{Z}) = H^0(G, A)$, and can be proven abstractly given certain assumptions (cf. (7.3)).

§ 3. The Exact Cohomology Sequence

After having introduced the cohomology groups $H^q(G, A)$, we now want to study how these groups behave in case either the module A or the group G changes. We will discuss the first case in this section.

If A and B are two G-modules and

$$f : A \longrightarrow B$$

is a G-homomorphism, then f canonically induces a homomorphism

$$\bar{f}_q : H^q(G, A) \longrightarrow H^q(G, B),$$

which arises in the following way. From the map

$$x(\sigma_1, \ldots, \sigma_q) \longmapsto fx(\sigma_1, \ldots, \sigma_q)$$

we get a homomorphism

$$f_q : A_q \longrightarrow B_q$$

between the groups of cochains A_q of A and B_q of B with the property that

$$\partial_{q+1} \circ f_q = f_{q+1} \circ \partial_{q+1}.$$

Therefore these maps fit into the infinite commutative diagram

$$
\begin{array}{ccccccc}
\cdots & \longrightarrow & A_q & \xrightarrow{\partial_{q+1}} & A_{q+1} & \longrightarrow & \cdots \\
 & & \Big\downarrow{f_q} & & \Big\downarrow{f_{q+1}} & & \\
\cdots & \longrightarrow & B_q & \xrightarrow{\partial_{q+1}} & B_{q+1} & \longrightarrow & \cdots
\end{array}
$$

which means precisely that

$$x(\sigma_1, \ldots, \sigma_q) \longmapsto fx(\sigma_1, \ldots, \sigma_q)$$

takes cocycles to cocycles and coboundaries to coboundaries. It follows that the homomorphism $f_q : A_q \to B_q$ induces canonically a homomorphism

$$\bar{f}_q : H^q(G, A) \longrightarrow H^q(G, B).$$

If $c \in H^q(G, A)$, the image $\bar{f}_q c$ is obtained by choosing a cocycle x from the class c, and taking the cohomology class of the cocycle fx of the module B.

We thus have a very simple explicit description of the homomorphism $\bar{f}_q$. This is an advantage that does not occur very often in cohomology theory. In fact, for many cohomological maps one knows only of their existence and their functorial properties, without having an explicit description. Nevertheless, it is equally significant that in the entire theory one almost always only works with these functorial properties, and an explicit description of the maps is required only in a few cases.

A first example of this typical situation is provided by the for the entire theory fundamental **connecting homomorphism** δ. Although this map is still given explicitly, its definition does not exactly leave the impression of great clarity and immediacy.

(3.1) Proposition. *If*

$$0 \longrightarrow A \xrightarrow{\;i\;} B \xrightarrow{\;j\;} C \longrightarrow 0$$

is an exact sequence of G-modules and G-homomorphisms, then there exists a canonical homomorphism

$$\delta_q : H^q(G, C) \longrightarrow H^{q+1}(G, A).$$

The map δ_q is called the **connecting homomorphism** *or also the* δ-*homomorphism.*

For the construction of δ_q consider the following commutative diagram

$$
\begin{array}{ccccccccc}
0 & \longrightarrow & A_{q-1} & \xrightarrow{\;i\;} & B_{q-1} & \xrightarrow{\;j\;} & C_{q-1} & \longrightarrow & 0 \\
& & \downarrow{\scriptstyle\partial} & & \downarrow{\scriptstyle\partial} & & \downarrow{\scriptstyle\partial} & & \\
0 & \longrightarrow & A_q & \xrightarrow{\;i\;} & B_q & \xrightarrow{\;j\;} & C_q & \longrightarrow & 0 \\
& & \downarrow{\scriptstyle\partial} & & \downarrow{\scriptstyle\partial} & & \downarrow{\scriptstyle\partial} & & \\
0 & \longrightarrow & A_{q+1} & \xrightarrow{\;i\;} & B_{q+1} & \xrightarrow{\;j\;} & C_{q+1} & \longrightarrow & 0.
\end{array}
$$

(For simplicity we have omitted the indices on the maps i, j, ∂.) The rows in this diagram result from applying the functor $\mathrm{Hom}_G(X_i, \ )$ $(i = q{-}1, q, q{+}1)$ to the exact sequence $0 \to A \to B \to C \to 0$; since the G-modules X_i are free (cf. (1.6)), it follows that these rows are exact.

We denote by a_q, b_q, c_q elements of the groups of cochains A_q, B_q, C_q, and write $\bar{a}_q, \bar{b}_q, \bar{c}_q$ for their images in the cohomology groups $H^q(G, A)$, $H^q(G, B)$, $H^q(G, C)$.

Assume now $\bar{c}_q \in H^q(G, C)$, so that $\partial c_q = 0$. We choose a b_q such that

$$c_q = jb_q.$$

We have $j\partial b_q = \partial j b_q = \partial c_q = 0$, and thus $\partial b_q \in \ker j_{q+1}$. Hence there exists a a_{q+1} with $\partial b_q = i a_{q+1}$. Because $i\partial a_{q+1} = \partial i a_{q+1} = \partial\partial b_q = 0$, i.e., $\partial a_{q+1} = 0$, we see that a_{q+1} is a $(q+1)$-cocycle of A. Now we set

$$\delta_q \bar{c}_q = \bar{a}_{q+1}.$$

This definition depends, of course, on the choice of the representative c_q of $\bar{c}_q$ and its preimage b_q. However, if we choose another representative c'_q with preimage b'_q (i.e., $j b'_q = c'_q$) and let $\overline{a'}_{q+1}$ be the resulting class, we have

$$\bar{c}_q = \bar{c'}_q \Rightarrow c_q - c'_q = \partial c_{q-1} \text{ for a } c_{q-1} \Rightarrow c_q - c'_q = \partial j b_{q-1} \text{ for a } b_{q-1}$$
$$\Rightarrow j b_q - j b'_q = j \partial b_{q-1} \Rightarrow b_q - b'_q - \partial b_{q-1} \in \ker j_q = \operatorname{im} i_q \Rightarrow i a_q = b_q - b'_q - \partial b_{q-1}$$
$$\text{for an } a_q \Rightarrow \partial i a_q = \partial b_q - \partial b'_q \Rightarrow i \partial a_q = i a_{q+1} - i a'_{q+1} \Rightarrow \partial a_q = a_{q+1} - a'_{q+1} \Rightarrow$$
$$\bar{a}_{q+1} = \overline{a'}_{q+1}.$$

Therefore δ_q is well defined; it is immediate that δ is a homomorphism.

As already mentioned, it is not necessary to recall the explicit definition of the connecting homomorphism δ every time this map comes up. Once we have proved the fundamental property of this map, its explicit definition is used only occasionally. This fundamental and most important property is given by the following theorem, which can be considered the main theorem of cohomology theory.

(3.2) Theorem. *Let*

$$0 \longrightarrow A \xrightarrow{i} B \xrightarrow{j} C \longrightarrow 0$$

be an exact sequence of G-modules and G-homomorphisms. Then the induced infinite sequence

$$\cdots \longrightarrow H^q(G,A) \xrightarrow{\bar{i}_q} H^q(G,B) \xrightarrow{\bar{j}_q} H^q(G,C) \xrightarrow{\delta_q} H^{q+1}(G,A) \longrightarrow \cdots$$

*is also exact. It is called the **exact cohomology sequence**.*

Proof. The homomorphisms $\bar{i}_q$, $\bar{j}_q$ and δ_q are respectively induced by:

$$a_q \mapsto i a_q, \quad b_q \mapsto j b_q, \quad c_q \mapsto a_{q+1},$$

where $c_q = j b_q$ and $\partial b_q = i a_{q+1}$. It follows that

$$\bar{j}_q \circ \bar{i}_q = 0, \text{ because } a_q \mapsto i a_q \mapsto j i a_q = 0,$$
$$\delta_q \circ \bar{j}_q = 0, \text{ because } b_q \mapsto j b_q \mapsto a_{q+1} = 0 \text{ (we have } i a_{q+1} = \partial b_q = 0),$$
$$\bar{i}_{q+1} \circ \delta_q = 0, \text{ because } c_q \mapsto a_{q+1} \mapsto i a_{q+1} = \partial b_q \in \partial B_q.$$

From this we obtain the inclusions

$$\operatorname{im} \bar{i}_q \subseteq \ker \bar{j}_q, \ \operatorname{im} \bar{j}_q \subseteq \ker \delta_q, \ \operatorname{im} \delta_q \subseteq \ker \bar{i}_{q+1}.$$

Let $\bar{b}_q \in \ker \bar{j}_q$ so that $jb_q = \partial c_{q-1}$ for a c_{q-1}. If we choose a b_{q-1} with $jb_{q-1} = c_{q-1}$, it follows that $j(b_q - \partial b_{q-1}) = 0$. Hence we may assume from the beginning that the representative b_q of $\bar{b}_q$ satisfies $jb_q = 0$. Then there exists an a_q with $b_q = ia_q$, and this a_q is a cocycle since $i\partial a_q = \partial b_q = 0$. Hence $\bar{b}_q = \bar{i}_q \bar{a}_q \in \operatorname{im} \bar{i}_q$, which proves the inclusion $\operatorname{im} \bar{i}_q \supseteq \ker \bar{j}_q$.

Let $\bar{c}_q \in \ker \delta_q$. By definition of δ_q, there are elements a_{q+1} and b_q such that $\delta_q \bar{c}_q = \bar{a}_{q+1} = 0$, $ia_{q+1} = \partial b_q$, and $c_q = jb_q$. Because $\bar{a}_{q+1} = 0$ we have $a_{q+1} = \partial a_q$, which implies that $\partial(b_q - ia_q) = 0$ and $c_q = j(b_q - ia_q)$. From this we obtain $\bar{c}_q = \bar{j}_q(\overline{b_q - ia_q})$, which shows that $\operatorname{im} \bar{j}_q \supseteq \ker \delta_q$.

Let $\bar{a}_{q+1} \in \ker \bar{i}_{q+1}$ so that $ia_{q+1} = \partial b_q$ for some b_q. If we let $c_q = jb_q$, then $\partial c_q = \partial j b_q = j \partial b_q = j i a_{q+1} = 0$. This shows that c_q is a cocycle and implies $\bar{a}_{q+1} = \delta_q \bar{c}_q \in \operatorname{im} \delta_q$. It follows that $\operatorname{im} \delta_q \supseteq \ker \bar{i}_{q+1}$, which completes the proof of the exactness of the cohomology sequence.

When we introduced the cohomology groups we already mentioned that working with a complete free resolution of G leads to a unification of homology and cohomology groups. The essential aspect here is not so much to have a unified notation but rather the existence of an exact sequence ranging from $-\infty$ to $+\infty$ that involves both the homology as well as the cohomology groups.

Theorem (3.2) is applied mostly frequently in the following form: If an arbitrary term in the exact cohomology sequence

$$\cdots \longrightarrow H^q(G, A) \longrightarrow H^q(G, B) \longrightarrow H^q(G, C) \longrightarrow H^{q+1}(G, A) \longrightarrow \cdots$$

vanishes, then the preceding map is surjective and the subsequent map is injective. This type of argument will allow us frequently to prove important isomorphism results. We state this in the following corollary.

(3.3) Corollary. *Let*

$$0 \longrightarrow A \xrightarrow{i} B \xrightarrow{j} C \longrightarrow 0$$

be an exact sequence of G-modules. If $H^q(G, A) = 0$ for all q, then $\bar{j}_q : H^q(G, B) \longrightarrow H^q(G, C)$ is an isomorphism. Similarly, if $H^q(G, B) = 0$ (resp. $H^q(G, C) = 0$) for all q, then $\delta_q : H^q(G, C) \longrightarrow H^{q+1}(G, A)$ (resp. $\bar{i}_q : H^q(G, A) \longrightarrow H^q(G, B)$) is an isomorphism.

Because of this corollary the G-modules which have only trivial cohomology groups play a distinguished role.

We continue our discussion of the connecting map $A^G \xrightarrow{\delta} H^1(G, A)$ from p. 17 and show that the cohomology sequence (3.2) induces an exact sequence which terminates on the left.

(3.4) Theorem. *Let*

$$0 \longrightarrow A \xrightarrow{i} B \xrightarrow{j} C \longrightarrow 0$$

be an exact sequence of G-modules. Then the following sequence is exact:

$$0 \longrightarrow A^G \xrightarrow{i} B^G \xrightarrow{j} C^G \xrightarrow{\delta} H^1(G,A) \xrightarrow{\bar{i}_1} H^1(G,B) \xrightarrow{\bar{j}_1} \cdots .$$

Proof.　The homomorphism $C^G \xrightarrow{\delta} H^1(G,A)$ is defined as the composition

$$C^G \longrightarrow C^G/N_G C = H^0(G,C) \xrightarrow{\delta_0} H^1(G,A);$$

hence it suffices to show exactness at the term C^G. To show $\operatorname{im} j \subseteq \ker \delta$ assume $c \in \operatorname{im} j \subseteq C^G$. Then $c = jb$ with $b \in B^G$, and the claim follows from

$$\delta c = \delta_0(c + N_G C) = \delta_0(jb + N_G C) = \delta_0 \bar{j}_0(b + N_G B) = 0.$$

For the opposite inclusion $\ker \delta \subseteq \operatorname{im} j$, consider $c \in \ker \delta$. Then $c \in C^G$ and $\delta c = \delta_0(c + N_G C) = 0$. It follows from (3.2) that we have the identity

$$c + N_G C = \bar{j}_0(b + N_G B) = jb + N_G C,$$

hence $c = jb + N_G c'$. If we choose a $b' \in B$ with $jb' = c'$, it follows that $c = jb + N_G(jb') = jb + jN_G b' \in jB^G$, which proves our claim.

Because of the exact sequence (3.4) the fixed modules A^G, B^G, and C^G are often defined as the zeroth cohomology groups; in particular, in case one is only interested in cohomology groups of positive dimension.

In the following we consider various compatibility properties of the connecting homomorphism δ.

(3.5) Proposition. *If*

$$
\begin{array}{ccccccccc}
0 & \longrightarrow & A & \xrightarrow{i} & B & \xrightarrow{j} & C & \longrightarrow & 0 \\
  &   & \downarrow{f} &   & \downarrow{g} &   & \downarrow{h} &   &   \\
0 & \longrightarrow & A' & \xrightarrow{i'} & B' & \xrightarrow{j'} & C' & \longrightarrow & 0
\end{array}
$$

is a commutative diagram of G-modules and G-homomorphisms with exact rows, then

$$\bar{f}_{q+1} \circ \delta_q = \delta_q \circ \bar{h}_q;$$

in other words, the following diagram is commutative:

$$
\begin{array}{ccc}
H^q(G,C) & \xrightarrow{\delta_q} & H^{q+1}(G,A) \\
\downarrow{\bar{h}_q} &   & \downarrow{\bar{f}_{q+1}} \\
H^q(G,C') & \xrightarrow{\delta_q} & H^{q+1}(G,A').
\end{array}
$$

This follows almost immediately from the definition of δ_q. Let $\bar{c}_q \in H^q(G,C)$. If we choose b_q and a_{q+1} such that $c_q = jb_q$ and $ia_{q+1} = \partial b_q$, then $\delta_q \bar{c}_q = \bar{a}_{q+1}$, so that $(\bar{f}_{q+1} \circ \delta_q)\bar{c}_q = \bar{f}_{q+1}(\bar{a}_{q+1}) = \overline{fa}_{q+1}$. If we set $c'_q = hc_q$, $b'_q = gb_q$ and $a'_{q+1} = fa_{q+1}$, it follows that $c'_q = j'b'_q$ and $\partial b'_q = i'a'_{q+1}$, from which we obtain $(\delta_q \circ \bar{h}_q)\bar{c}_q = \delta_q \bar{c}'_q = \bar{a}'_{q+1} = \overline{fa}_{q+1} = \bar{f}_{q+1} \circ \delta_q)\bar{c}_q$, i.e., $\bar{f}_{q+1} \circ \delta_q = \delta_q \circ \bar{h}_q$.

The connecting homomorphism δ is "anticommutative":

(3.6) Theorem. *Assume the diagram of G-modules and G-homomorphisms*

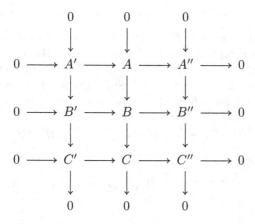

is commutative with exact rows and columns. Then the diagram

$$
\begin{array}{ccc}
H^{q-1}(G,C'') & \xrightarrow{\ \delta\ } & H^q(G,C') \\
\downarrow{\scriptstyle \delta} & & \downarrow{\scriptstyle -\delta} \\
H^q(G,A'') & \xrightarrow{\ \delta\ } & H^{q+1}(G,A')
\end{array}
$$

commutes.

Proof. Let D be the kernel of the composite map $B \to C''$; thus the sequence

$$0 \longrightarrow D \longrightarrow B \longrightarrow C'' \longrightarrow 0$$

is exact. We define G-homomorphisms

$i : A' \to A \oplus B'$ by $i(a') = (a, b')$, where a (resp. b') is the image of a' in A (resp. of a' in B'),

$j : A \oplus B' \to D$ by $j(a, b') = d_1 - d_2$, where d_1 (resp. d_2) is the image of a (resp. of b') in $D \subseteq B$.

It is easy to verify that with these definitions the sequence

$$0 \longrightarrow A' \xrightarrow{\ i\ } A \oplus B' \xrightarrow{\ j\ } D \longrightarrow 0$$

is exact, and that the diagram

$$
\begin{array}{ccccccccc}
A' & \longrightarrow & A & \longrightarrow & A'' & \longrightarrow & B'' & \longrightarrow & C'' \\
\Vert \text{id} & & \uparrow (\text{id},0) & & \uparrow & & \uparrow & & \Vert \text{id} \\
A' & \xrightarrow{i} & A \oplus B' & \xrightarrow{j} & D & \longrightarrow & B & \longrightarrow & C'' \\
\Vert -\text{id} & & \downarrow (0,-\text{id}) & & \downarrow & & \downarrow & & \Vert \text{id} \\
A' & \longrightarrow & B' & \longrightarrow & C' & \longrightarrow & C & \longrightarrow & C''
\end{array}
$$

commutes. Because $\mathrm{im}\,(D \to B'') \subseteq \mathrm{im}\,(A'' \to B'')$ and $A'' \to B''$ is injective, there is a G-homomorphism $D \to A''$ which extends the above diagram. Similarly, since $\mathrm{im}\,(D \to C) \subseteq \mathrm{im}\,(C' \to C)$ and $C' \to C$ is injective, there is an analogous G-homomorphism $D \to C'$. Since the resulting extended diagram is commutative, it follows from (3.5) that the following diagram

$$
\begin{array}{ccccc}
H^{q-1}(G,C'') & \xrightarrow{\ \delta\ } & H^q(G,A'') & \xrightarrow{\ \delta\ } & H^{q+1}(G,A') \\
\Vert \text{id} & & \uparrow & & \Vert \text{id} \\
H^{q-1}(G,C'') & \xrightarrow{\ \delta\ } & H^q(G,D) & \xrightarrow{\ \delta\ } & H^{q+1}(G,A') \\
\Vert \text{id} & & \downarrow & & \Vert -\text{id} \\
H^{q-1}(G,C'') & \xrightarrow{\ \delta\ } & H^q(G,C') & \xrightarrow{\ \delta\ } & H^{q+1}(G,A'),
\end{array}
$$

commutes as well, which immediately implies the theorem.

(3.7) Proposition. *Let* $\{A_\iota \mid \iota \in I\}$ *be a family of G-modules. Then*

$$H^q\Big(G, \bigoplus_\iota A_\iota\Big) \cong \bigoplus_\iota H^q(G, A_\iota).$$

Proof. If we set $A = \bigoplus_\iota A_\iota$, we have from (1.5) a canonical isomorphism

$$A_q = \mathrm{Hom}_G(X_q, A) \cong \bigoplus_\iota \mathrm{Hom}_G(X_q, A_\iota) = \bigoplus_\iota (A_\iota)_q,$$

and the proposition follows from the infinite commutative diagram

$$
\begin{array}{ccccc}
\cdots \longrightarrow & A_{q-1} & \xrightarrow{\ \partial\ } & A_q & \longrightarrow \cdots \\
& \downarrow \wr & & \downarrow \wr & \\
\cdots \longrightarrow & \bigoplus_\iota (A_\iota)_{q-1} & \xrightarrow{\ \partial\ } & \bigoplus_\iota (A_\iota)_q & \longrightarrow \cdots .
\end{array}
$$

The same reasoning applies to the direct product $\prod_\iota A_\iota$ in place of the direct sum $\bigoplus_\iota A_\iota$, using the evident isomorphism

$$\left(\prod_\iota A_\iota\right)_q = \operatorname{Hom}_G\left(X_q, \prod_\iota A_\iota\right) \cong \prod_\iota \operatorname{Hom}_G(X_q, A_\iota) = \prod_\iota (A_\iota)_q.$$

Therefore we also have

(3.8) Proposition. $H^q(G, \prod_\iota A_\iota) \cong \prod_\iota H^q(G, A_\iota)$

The G-induced modules. We explained already in (3.3) that the exact cohomology sequence yields isomorphism theorems in case the underlying exact sequence contains a G-module with only trivial cohomology groups. A particular class of such G-modules are the **G-induced** modules, which we will make use of in many of the proofs and definitions below.

(3.9) Definition. *A G-module A is called **G-induced** if it can be represented as a direct sum*

$$A = \bigoplus_{\sigma \in G} \sigma D$$

with a subgroup $D \subseteq A$.

In particular, the G-module $\mathbb{Z}[G] = \bigoplus_{\sigma \in G} \sigma(\mathbb{Z}\cdot 1)$ is G-induced, and it is clear that the G-induced modules are represented simply as the tensor products

$$\mathbb{Z}[G] \otimes D$$

with arbitrary abelian groups D. In fact, if we consider D as a trivial G-module, we have the G-isomorphism

$$\mathbb{Z}[G] \otimes D = \left(\bigoplus_{\sigma \in G} \mathbb{Z}\sigma\right) \otimes D = \bigoplus_{\sigma \in G} \mathbb{Z}(\sigma \otimes D) = \bigoplus_{\sigma \in G} \sigma(\mathbb{Z} \otimes D).$$

More generally, we have

(3.10) Proposition. *Let X be a G-induced module and A an arbitrary G-module. Then the module $X \otimes A$ is also G-induced.*

By assumption $X = \bigoplus_{\sigma \in G} \sigma D$, which implies

$$X \otimes A = \left(\bigoplus_{\sigma \in G} \sigma D\right) \otimes A \cong \bigoplus_{\sigma \in G} (\sigma D) \otimes (\sigma A) \cong \bigoplus_{\sigma \in G} \sigma(D \otimes A).$$

(3.11) Proposition. *Let A be a G-induced module, and g a subgroup of G. Then A is a g-induced g-module. If g is normal in G, then A^g is a G/g-induced G/g-module.*

Proof. Let $A = \bigoplus_{\sigma \in G} \sigma D$. Then we can write

$$A = \bigoplus_{\sigma \in g} \bigoplus_\tau \sigma\tau D = \bigoplus_{\sigma \in g} \sigma\left(\bigoplus_\tau \tau D\right),$$

where τ ranges over a system of right coset representatives of G with respect to g. Hence A is g-induced.

Assume g is normal in G. We show the G/g-module A^g has the representation

$$A^g = \bigoplus_{\tau \in G/g} \tau N_g D.$$

Because of the direct sum decomposition $A = \bigoplus_{\sigma \in G} \sigma D$, the sum on the right side of the above identity is obviously direct. Since $N_g D \subseteq A^g$, it is also contained in A^q. Conversely, suppose $a \in A^g$. The element a has a unique representation as $a = \sum_{\tau \in G} \tau d_\tau$ with $d_\tau \in D$. If $\sigma \in g$, it follows that

$$a = \sigma a = \sum_{\tau \in G} \sigma\tau d_\tau = \sum_{\tau \in G} \sigma\tau d_{\sigma\tau},$$

and by uniqueness that $d_\tau = d_{\sigma\tau}$. From this we obtain the representation

$$a = \sum_\tau \sum_{\sigma \in g} \tau\sigma d_{\tau\sigma} = \sum_\tau \tau\left(\sum_{\sigma \in g} \sigma d_\tau\right) = \sum_\tau \tau N_g(d_\tau),$$

where τ ranges over a system of left coset representatives of G/g. This proves that A^g is G/g-induced.

(3.12) **Definition.** *We say that a G-module A has* **trivial cohomology** *if*

$$H^q(g, A) = 0$$

for all q and all subgroups $g \subseteq G$.

The next theorem is crucial:

(3.13) **Theorem.** *Every G-induced module A has trivial cohomology.*

Proof. Because of (3.11) it suffices to prove $H^q(G, A) = 0$, i.e., the sequence

$$\cdots \longrightarrow \mathrm{Hom}_G(X_q, A) \xrightarrow{\partial} \mathrm{Hom}_G(X_{q+1}, A) \longrightarrow \cdots$$

is exact. Let $A = \bigoplus_{\sigma \in G} \sigma D$, and let $\pi : A \to D$ is the natural projection of A onto D. Then the map $f \mapsto \pi \circ f$ induces an obviously bijective homomorphism

$$\mathrm{Hom}_G(X_q, A) \longrightarrow \mathrm{Hom}(X_q, D),$$

and we may identify $\mathrm{Hom}_G(X_q, A)$ with $\mathrm{Hom}(X_q, D)$. Therefore it suffices to consider the sequence

$$\cdots \longrightarrow \mathrm{Hom}(X_q, D) \longrightarrow \mathrm{Hom}(X_{q+1}, D) \longrightarrow \cdots,$$

which is exact by (1.7).

Because G-induced modules have trivial cohomology, we obtain from (3.3) a very important potential application of G-induced modules, using the fact that every G-module A can be viewed as a submodule as well as a factor module of a G-induced module, as we explain next.

As before, we denote by I_G the augmentation ideal of $\mathbb{Z}[G]$, and by J_G the factor module $J_G = \mathbb{Z}[G]/\mathbb{Z} \cdot N_G$. By (1.2) all terms in the exact sequences

$$0 \longrightarrow I_G \longrightarrow \mathbb{Z}[G] \overset{\varepsilon}{\longrightarrow} \mathbb{Z} \longrightarrow 0 \,,$$
$$0 \longrightarrow \mathbb{Z} \overset{\mu}{\longrightarrow} \mathbb{Z}[G] \longrightarrow J_G \longrightarrow 0 \,.$$

consist of free $\mathbb{Z}$-modules. From (1.8) we immediately conclude

(3.14) Proposition. *For all G-modules A we have the exact sequences*

$$0 \longrightarrow I_G \otimes A \longrightarrow \mathbb{Z}[G] \otimes A \longrightarrow \quad A \quad \longrightarrow 0 \,,$$
$$0 \longrightarrow \quad A \quad \longrightarrow \mathbb{Z}[G] \otimes A \longrightarrow J_G \otimes A \longrightarrow 0 \,.$$

Since the G-module $\mathbb{Z}[G] \otimes A$ is G-induced by (3.10), these exact sequences allow us to view A as a submodule as well as a factor module of a G-induced module.

Because $\mathbb{Z}[G] \otimes A$ is cohomologically trivial, applying the exact cohomology sequence to the exact sequences from Proposition (3.14) yields **isomorphisms**

$$\delta : \ H^{q-1}(g, A^1) \longrightarrow H^q(g, A) \qquad \text{with } A^1 = J_G \otimes A,$$
$$\delta^{-1} : H^{q+1}(g, A^{-1}) \longrightarrow H^q(g, A) \qquad \text{with } A^{-1} = I_G \otimes A$$

for every q and every subgroup $g \subseteq G$; this uses (3.3). We want to iterate this:

For every integer $m \in \mathbb{Z}$ set

$$A^m = \underbrace{J_G \otimes \cdots \otimes J_G}_{m\text{-times}} \otimes A, \quad \text{if } m \geq 0,$$

$$A^m = \underbrace{I_G \otimes \cdots \otimes I_G}_{|m|\text{-times}} \otimes A, \quad \text{if } m \leq 0.$$

From the composition of the isomorphism δ (resp. δ^{-1}) we obtain maps

$$H^{q-m}(g, A^m) \longrightarrow H^{q-(m-1)}(g, A^{m-1}) \longrightarrow \cdots \longrightarrow H^q(g, A)$$

and thus isomorphisms

$$\delta^m : H^{q-m}(g, A^m) \longrightarrow H^q(g, A) \qquad (m \in \mathbb{Z}).$$

This shows:

(3.15) Proposition. *Let A be a G-module. There are G-modules*

$$A^m = J_G \otimes \cdots \otimes J_G \otimes A \qquad (m \geq 0), \quad resp.$$
$$A^m = I_G \otimes \cdots \otimes I_G \otimes A \qquad (m \leq 0)$$

with the property that the m-fold composition of the connecting homomorphism δ induces for every q and every subgroup $g \subseteq G$ an isomorphism

$$\delta^m : H^{q-m}(g, A^m) \longrightarrow H^q(g, A) \qquad (m \in \mathbb{Z})$$

We will frequently use the isomorphism $H^q(g, A) \cong H^{q-m}(g, A^m)$ to deduce from statements about cohomology groups of dimension q analogous statements for a higher or lower dimension. In particular, this technique will allow us to reduce many definitions and proofs to the case of zero-dimensional cohomology groups, which we understand much better that the higher-dimensional analogues. This method is called the **method of dimension shifting**[10].

The following theorem gives a first example of the usefulness of this method:

(3.16) Theorem. *Let A be a G-module. Then $H^q(G, A)$ is a torsion groups and the orders of the elements in $H^q(G, A)$ divide the order n of the group G*

$$n \cdot H^q(G, A) = 0.$$

Proof. If $q = 0$, then $n \cdot H^0(G, A) = 0$, because $H^0(G, A) = A^G/N_G A$ and $na = N_G a$ for all $a \in A^G$. The general case follows from this and the isomorphism $H^q(G, A) \cong H^0(G, A^q)$.

(3.17) Corollary. *A uniquely divisible*[11] *G-module A has trivial cohomology.*

If A is divisible, the map 'multiplication by n', i.e., $n \cdot \mathrm{id} : A \to A$, is bijective for every natural number n, and therefore induces isomorphisms

$$n \cdot \mathrm{id} : H^q(g, A) \longrightarrow H^q(g, A) \qquad (g \subseteq G).$$

If $n = |G|$, it follows from (3.16) that $H^q(g, A) = n \cdot H^q(g, A) = 0$.

---

[10] Cohomology theory can be based directly on this principle. In fact, by (3.15)

$$H^q(G, A) \cong H^0(G, A^q),$$

where A^q is given canonically by A: $A^q = J_G \otimes \ldots \otimes J_G \otimes A$ for $q \geq 0$, resp. $A^q = I_G \otimes \ldots \otimes I_G \otimes A$ for $q \leq 0$. Therefore one may define the cohomology groups of the G-module A from the beginning as the quotient group

$$H^q(G, A) = (A^q)^G/N_G A^q.$$

For cohomology theory developed along these lines, see C. CHEVALLEY [12].

[11] An abelian group A is said to be uniquely divisible if for every $a \in A$ and every natural number n the equation $nx = a$ has a unique solution $x \in A$.

In particular, the G-module $\mathbb{Q}$ (on which the group G always acts trivially) has trivial cohomology. From the cohomology sequence associated with the exact sequence

$$0 \longrightarrow \mathbb{Z} \longrightarrow \mathbb{Q} \longrightarrow \mathbb{Q}/\mathbb{Z} \longrightarrow 0$$

we obtain

(3.18) Corollary. *There is a canonical isomorphism*

$$H^2(G, \mathbb{Z}) \cong H^1(G, \mathbb{Q}/\mathbb{Z}) = \mathrm{Hom}(G, \mathbb{Q}/\mathbb{Z}) = \chi(G).$$

The group $\chi(G) = \mathrm{Hom}(G, \mathbb{Q}/\mathbb{Z})$ is called the **character group** of G.

We end this section with the computation of the group $H^{-2}(G, \mathbb{Z})$, which plays an important role in class field theory. We denote the commutator subgroup of G by G', and its abelianization by $G^{\mathrm{ab}} = G/G'$.

(3.19) Theorem. *There is a canonical isomorphism $H^{-2}(G, \mathbb{Z}) \cong G^{\mathrm{ab}}$.*

Proof. Since $\mathbb{Z}[G]$ is a G-induced module, it has trivial cohomology, and we obtain from the exact cohomology sequence associated with

$$0 \longrightarrow I_G \longrightarrow \mathbb{Z}[G] \overset{\varepsilon}{\longrightarrow} \mathbb{Z} \longrightarrow 0$$

the isomorphism

$$\delta : H^{-2}(G, \mathbb{Z}) \longrightarrow H^{-1}(G, I_G)$$

Since $H^{-1}(G, I_G) = I_G/I_G^2$ it suffices to produce an isomorphism $G/G' \cong I_G/I_G^2$. (Note that G is written multiplicatively and I_G is written additively.) For this we consider the map

$$G \longrightarrow I_G/I_G^2, \quad \sigma \longmapsto (\sigma - 1) + I_G^2.$$

Because $\sigma \cdot \tau - 1 = (\sigma - 1) + (\tau - 1) + (\sigma - 1) \cdot (\tau - 1)$, this map is a homomorphism. Since I_G/I_G^2 is abelian, the kernel of this homomorphism contains the commutator subgroup G', which implies that we have a homomorphism

$$\log : G/G' \longrightarrow I_G/I_G^2.$$

In order to show that the map log is bijective, we use that I_G is the free abelian group generated by $\sigma - 1$, where $\sigma \in G \smallsetminus \{1\}$. Hence setting

$$\sigma - 1 \longmapsto \sigma \cdot G'$$

defines an evidently surjective homomorphism from I_G to G/G'. Because

$$(\sigma - 1) \cdot (\tau - 1) = (\sigma\tau - 1) - (\sigma - 1) - (\tau - 1) \longmapsto \sigma\tau\sigma^{-1}\tau^{-1}G' = \bar{1},$$

the elements in I_G^2 lie in the kernel, so that we obtain a homomorphism

$$\exp : I_G/I_G^2 \longrightarrow G/G', \quad (\sigma - 1) + I_G^2 \longmapsto \sigma G'$$

with the property that $\log \circ \exp = \mathrm{id}$ and $\exp \circ \log = \mathrm{id}$. Therefore the map $\log : G/G' \to I_G/I_G^2$ is an isomorphism.

It is easy to see that for the G-module $\mathbb{Z}$ we have $H^{-1}(G, \mathbb{Z}) = {}_{N_G}\mathbb{Z}/I_G\mathbb{Z} = 0$, $H^0(G, \mathbb{Z}) = \mathbb{Z}/n\mathbb{Z}$, and $H^1(G, \mathbb{Z}) = \operatorname{Hom}(G, \mathbb{Z}) = 0$. Thus we have determined the cohomology groups $H^q(G, \mathbb{Z})$ of dimensions $q = -2, -1, 0, 1, 2$:

$$H^{-2}(G, \mathbb{Z}) \cong G^{\mathrm{ab}}, \ H^{-1}(G, \mathbb{Z}) = 0, \ H^0(G, \mathbb{Z}) = \mathbb{Z}/n\mathbb{Z},$$

$$H^1(G, \mathbb{Z}) = 0, \ H^2(G, \mathbb{Z}) = \chi(G).$$

We mention without proof that there is a canonical isomorphism (by duality)

$$H^{-q}(G, \mathbb{Z}) \cong \chi(H^q(G, \mathbb{Z})) \quad \text{for all } q > 0.$$

§ 4. Inflation, Restriction and Corestriction

In the previous section we studied the dependence of the cohomology groups $H^q(G, A)$ on the module A; now we consider the behavior of these groups in case the group G varies. This mainly concerns the following questions.

Let A be a G-module and let g be a subgroup of G. Then A is always a g-module and A^g is a G/g-module, provided g is normal in G. What are the relations among the cohomology groups

$$H^q(G/g, A^g), \ H^q(G, A) \text{ and } H^q(g, A) \, ?$$

We first restrict our considerations to the case of positive dimension $q \geq 1$.

If g is normal in G, we associate with every q-cochain

$$x : G/g \times \cdots \times G/g \longrightarrow A^g$$

a q-cochain

$$y : G \times \cdots \times G \longrightarrow A$$

by defining $y(\sigma_1, \ldots, \sigma_q) = x(\sigma_1 \cdot g, \ldots, \sigma_q \cdot g)$. We call this y the **inflation** of x and denote it by

$$y = \inf x.$$

It is easy to see that the map $x \mapsto \inf x$ is compatible with the coboundary operator ∂, i.e., $\partial_{q+1} \circ \inf = \inf \circ \partial_{q+1}$. Hence cocycles are mapped to cocycles and coboundaries to coboundaries, and we obtain a map on cohomology:

(4.1) Definition. *Let A be a G-module and g a normal subgroup of G. The homomorphism*

$$\inf_q : H^q(G/g, A^g) \longrightarrow H^q(G, A), \quad q \geq 1,$$

induced by the homomorphism from the q-th group of cochains of the G/g-module A^g to the q-th group of cochains of the G-module A is called **inflation**.

Along with inflation we obtain another cohomological map by associating with every q-cochain

$$x : G \times \cdots \times G \longrightarrow A$$

its restriction

$$y : g \times \cdots \times g \longrightarrow A$$

from $G \times \cdots \times G$ to $g \times \cdots \times g$. We call this q-cochain y the **restriction** of x and denote it by

$$y = \operatorname{res} x.$$

The crucial point here is again that the cochain homomorphism res commutes with the operator ∂, i.e., $\partial_{q+1} \circ \operatorname{res} = \operatorname{res} \circ \partial_{q+1}$, hence there is an induced map on cohomology groups:

(4.2) Definition. *Let A be a G-module and g a subgroup of G. The homomorphism*

$$\operatorname{res}_q : H^q(G, A) \longrightarrow H^q(g, A), \quad q \geq 1,$$

induced by the restriction of the cochains of the G-module A to the group g is called **restriction**.

To make sure the cohomological maps defined above fit into the general theory, we have to verify that they are compatible with the canonical homomorphisms already given. This is the content of the following propositions.

(4.3) Proposition. *Let A and B be two G-modules, g a normal subgroup of G, and*

$$f : A \longrightarrow B$$

a G-homomorphism. Then the following diagrams

$$
\begin{array}{ccc}
H^q(G/g, A^g) & \xrightarrow{\ \bar{f}\ } & H^q(G/g, B^g) \\
\downarrow{\inf_q} & & \downarrow{\inf_q} \\
H^q(G, A) & \xrightarrow{\ \bar{f}\ } & H^q(G, B),
\end{array}
\qquad
\begin{array}{ccc}
H^q(G, A) & \xrightarrow{\ \bar{f}\ } & H^q(G, B) \\
\downarrow{\operatorname{res}_q} & & \downarrow{\operatorname{res}_q} \\
H^q(g, A) & \xrightarrow{\ \bar{f}\ } & H^q(g, B)
\end{array}
$$

are commutative. In the second diagram the normality of g in G is not needed.

Note here that the G-homomorphism $f : A \to B$ induces a G/g-homomorphism $f : A^g \to B^g$ as well as a g-homomorphism $f : A \to B$.

(4.4) Proposition. *Let*

$$0 \longrightarrow A \longrightarrow B \longrightarrow C \longrightarrow 0$$

be an exact sequence of G-modules and G-homomorphisms, and let g be a normal subgroup of G. If the sequence

$$0 \longrightarrow A^g \longrightarrow B^g \longrightarrow C^g \longrightarrow 0$$

is also exact, then the diagram

$$H^q(G/g, C^g) \xrightarrow{\delta} H^{q+1}(G/g, A^g)$$

$$\downarrow \text{inf}_q \qquad\qquad\qquad \downarrow \text{inf}_{q+1}$$

$$H^q(G, C) \xrightarrow{\delta} H^{q+1}(G, A)$$

commutes.

(4.5) Proposition. *Let*

$$0 \longrightarrow A \longrightarrow B \longrightarrow C \longrightarrow 0$$

be an exact sequence of G-modules and G-homomorphisms, and let g be a subgroup of G. Then the diagram

$$H^q(G, C) \xrightarrow{\delta} H^{q+1}(G, A)$$

$$\downarrow \text{res}_q \qquad\qquad\qquad \downarrow \text{res}_{q+1}$$

$$H^q(g, C) \xrightarrow{\delta} H^{q+1}(g, A)$$

commutes.

Propositions (4.3), (4.4) and (4.5) are easy to verify. The proof of the last two statements follows essentially from the fact that the inflation and restriction maps commute with the operator ∂, together with the definition of δ. We leave the details to the reader.

If we compose inflation and restriction, we obtain the following relation:

(4.6) Theorem. *Let A be a G-module and g a normal subgroup of G. Then*

$$0 \longrightarrow H^1(G/g, A^g) \xrightarrow{\text{inf}} H^1(G, A) \xrightarrow{\text{res}} H^1(g, A)$$

is exact.

Proof. To show the inflation map is injective, let $x : G/g \to A^g$ be a 1-cocycle whose inflation inf x is a 1-coboundary of the G-module A. Then

$$\text{inf } x(\sigma) = x(\sigma \cdot g) = \sigma a - a, \quad a \in A.$$

Hence we have for all $\tau \in g$ the equation $\sigma a - a = \sigma \tau a - a$, i.e., $a = \tau a$ which implies $a \in A^g$. Therefore $x(\sigma g) = \sigma \cdot ga - a$ is a 1-coboundary.

In order to prove the exactness at the term $H^1(G, A)$, consider a 1-cocyle $x : G/g \to A^g$ of A^g. If $\sigma \in g$, it follows that

$$\text{res} \circ \text{inf } x(\sigma) = \text{inf } x(\sigma) = x(\sigma g) = x(g) = x(\bar{1}).$$

But now $x(\bar{1}) = x(\bar{1} \cdot \bar{1}) = x(\bar{1}) + x(\bar{1}) = 0$ which implies res$\circ$inf $= 0$. Therefore

$$\text{im inf} \subseteq \text{ker res}.$$

Conversely, let $x : G \to A$ be a 1-cocycle of the G-module A whose restriction to g is a 1-coboundary of the g-module A:

$$x(\tau) = \tau a - a, \ a \in A, \ \text{for all } \tau \in g.$$

If we subtract from x the 1-coboundary $\rho : G \to A$, $\rho(\sigma) = \sigma a - a$, $\sigma \in G$, we obtain a 1-cocycle $x'(\sigma) = x(\sigma) - \rho(\sigma)$ in the same cohomology class with $x'(\tau) = 0$ for all $\tau \in g$. Then

$$x'(\sigma - \tau) = x'(\sigma) + \sigma x'(\tau) = x'(\sigma) \quad \text{for all } \tau \in g,$$

and, on the other hand,

$$x'(\tau \cdot \sigma) = x'(\tau) + \tau x'(\sigma) = \tau x'(\sigma) \quad \text{for all } \tau \in g.$$

If we now define $y : G/g \to A$ by $y(\sigma \cdot g) = x'(\sigma)$, we have $y(\sigma \cdot g) \in A^g$ because $y(\sigma \cdot g) = y(\tau \sigma \cdot g)$ for all $\tau \in g$, and y defines a 1-cocycle with inf $y = x'$. This proves that ker res $\subseteq$ im inf.

The analogue of Theorem (4.6) for higher dimensions holds only under certain conditions:

(4.7) Theorem. *Let A be a G-module and g a normal subgroup of G. If $H^i(g, A) = 0$ for $i = 1, \ldots, q - 1$ and $q \geq 1$, then the sequence*

$$0 \longrightarrow H^q(G/g, A^g) \xrightarrow{\text{inf}} H^q(G, A) \xrightarrow{\text{res}} H^q(g, A)$$

is exact.

We prove this by induction on the dimension q, using dimension shifting (cf. §3.) and Theorem (4.6) as the initial induction step. If we set $B = \mathbb{Z}[G] \otimes A$ and $C = J_G \otimes A$, we have from (3.14) the exact sequence

$$0 \longrightarrow A \longrightarrow B \longrightarrow C \longrightarrow 0.$$

Because $H^1(g, A) = 0$, it follows from Theorem (3.4) that the sequence

$$0 \longrightarrow A^g \longrightarrow B^g \longrightarrow C^g \longrightarrow 0$$

is also exact. Hence we have the following commutative diagram

$$
\begin{array}{ccccccc}
0 & \longrightarrow & H^{q-1}(G/g, C^g) & \xrightarrow{\text{inf}} & H^{q-1}(G, C) & \xrightarrow{\text{res}} & H^{q-1}(g, C) \\
& & \downarrow{\scriptstyle\delta} & & \downarrow{\scriptstyle\delta} & & \downarrow{\scriptstyle\delta} \\
0 & \longrightarrow & H^q(G/g, A^g) & \xrightarrow{\text{inf}} & H^q(G, A) & \xrightarrow{\text{res}} & H^q(g, A).
\end{array}
$$

Since B is G-induced and g-induced, and B^g is G/g-induced (cf. (3.10) and (3.11)), the connecting maps δ are isomorphisms (cf. (3.3)). It follows that

$$H^i(g, C) \cong H^{i+1}(g, A) = 0 \quad \text{for } i = 1, \ldots, q - 2.$$

Hence if we assume by induction that the upper sequence in the above diagram is exact, then this also holds for the lower sequence.

Of course, one will ask why we have introduced the maps inf and res only in case of positive dimensions $q \geq 1$ instead of defining such maps analogously using cochains for all negative dimensions as well. However, this is not possible. In fact, the crucial property of inflation and restriction is that by (4.4) and (4.5) respectively they are transformed into one another through dimension shifting by the operator δ, and one would want this property to hold for a general definition in all dimensions as well. Now for inflation it is necessary to restrict to $q \geq 1$; this follows essentially because if $0 \to A \to B \to C \to 0$ is an exact sequence of G-modules, then $0 \to A^g \to B^g \to C^g \to 0$ ($g \subseteq G$) is in general not exact, i.e., the first sequence induces a δ-homomorphism on cohomology groups, but the second does not.

The situation is different for restriction, however. In fact, this map can be extended to all dimensions $q \leq 0$. For example, if $q = 0$, one obtains from

$$a + N_G A \longmapsto a + N_g A, \quad a \in A^G \subseteq A^g,$$

a homomorphism

$$\mathrm{res}_0 : H^0(G, A) = A^G/N_G A \longrightarrow H^0(g, A) = A^g/N_g A,$$

with the property that Proposition (4.5) remains valid for $q = 0$. We pin this down in the following lemma:

(4.8) Lemma. *Let $0 \to A \xrightarrow{i} B \xrightarrow{j} C \to 0$ be an exact sequence of G-modules, and let g be a subgroup of G. Then the following diagram commutes*

$$
\begin{array}{ccc}
H^0(G, C) & \xrightarrow{\;\delta\;} & H^1(G, A) \\
\downarrow{\scriptstyle \mathrm{res}_0} & & \downarrow{\scriptstyle \mathrm{res}_1} \\
H^0(g, C) & \xrightarrow{\;\delta\;} & H^1(g, A).
\end{array}
$$

Proof. Let $c \in C^G$ be a 0-cocycle of the G-module C and $\bar{c} = c + N_G C$ its cohomology class. Then $\mathrm{res}_0 \bar{c} = c + N_g C$, i.e., c is also a 0-cocycle for the g-module C. If we choose $b \in B$ with $jb = c$, then $j\partial b = \partial c = 0$ implies that there exists a 1-cocycle $a_1 : G \to A$ such that $ia_1 = \partial b$. By definition of δ we have $\delta \bar{c} = \bar{a}_1$, and therefore $\delta \mathrm{res}_0 \bar{c} = \overline{\mathrm{res}_1 a_1} = \mathrm{res}_1 \bar{a}_1 = \mathrm{res}_1 \delta \bar{c}$.

Unfortunately, a similarly elementary definition of the restriction maps res_q in dimensions $q < 0$ cannot be given. Nevertheless, we will see that if such a restriction is specified for a single dimension, say $q = 0$, the compatibility condition in (4.5) uniquely determines all other restriction maps. This leads us toward an axiomatic approach to restriction as follows.

(4.9) Definition. *Let G be a finite group and g a subgroup of G. Then* **restriction** *is the uniquely determined family of homomorphisms*

$$\mathrm{res}_q : H^q(G, A) \longrightarrow H^q(g, A), \qquad q \in \mathbb{Z},$$

with the properties:

(i) *If $q = 0$, then*

$$\mathrm{res}_0 : H^0(G, A) \longrightarrow H^0(g, A), \quad a + N_G A \mapsto a + N_g A \quad (a \in A^G).$$

(ii) *For every exact sequence $0 \to A \to B \to C \to 0$ of G-modules and G-homomorphisms, the following diagram is commutative*

$$
\begin{array}{ccc}
H^q(G, C) & \xrightarrow{\ \delta\ } & H^{q+1}(G, A) \\
\downarrow{\scriptstyle \mathrm{res}_q} & & \downarrow{\scriptstyle \mathrm{res}_{q+1}} \\
H^q(g, C) & \xrightarrow{\ \delta\ } & H^{q+1}(g, A)
\end{array}
$$

The restrictions res_q are obtained from res_0 by dimension shifting as follows:

By (3.15) we have the isomorphisms

$$\delta^q : H^0(G, A^q) \longrightarrow H^q(G, A) \text{ and } \delta^q : H^0(g, A^q) \longrightarrow H^q(g, A),$$

given by the q-fold compositions of the connecting homomorphism δ. Condition (ii) now means that we have to define res_q by the commutative diagram

$$
\begin{array}{ccc}
H^0(G, A^q) & \xrightarrow{\ \delta^q\ } & H^q(G, A) \\
\downarrow{\scriptstyle \mathrm{res}_0} & & \downarrow{\scriptstyle \mathrm{res}_q} \\
H^0(g, A^q) & \xrightarrow{\ \delta^q\ } & H^q(g, A).
\end{array}
$$

This also shows uniqueness of the restriction maps. In particular, the restrictions res_q for $q \geq 0$ defined in this way coincide with those introduced earlier.

It remains to show that the homomorphisms res_q satisfy condition (ii). To this end we consider the following diagram

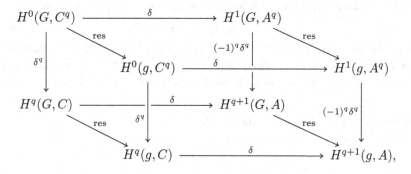

where we have used that from the sequence $0 \to A \to B \to C \to 0$ we obtain by induction using Proposition (1.2) and Lemma (1.9), the exact sequence

$$0 \longrightarrow A^q \longrightarrow B^q \longrightarrow C^q \longrightarrow 0.$$

In the above diagram the upper square commutes by (4.8). The commutativity of the two side diagrams follows immediately from the definition of the restriction maps by dimension shifting. The back and front diagrams are obtained by composing q squares of the type (3.6); hence by (3.6) they also commute. Therefore the commutativity of the upper square implies the commutativity of the lower square, which completes the proof.

Concerning an explicit description of the homomorphisms res_q for $q < 0$, i.e., the question of how the individual cocycles behave under these mappings, we remark that only by extensive calculations one can achieve some results, which in turn are hardly useful because they are far too technical. Nevertheless, our remarks on p. 21 about the general nature of cohomological methods also apply here. It is essentially the functorial properties of the restriction that come up, and only in small dimensions, where we have a concrete interpretation of the cohomology groups, we occasionally have to use an explicit description.

Using the isomorphism from Theorem (3.19), we point out a special case of a restriction which is important for class field theory:

(4.10) Definition. *Let $g \subseteq G$ be a subgroup. The homomorphism*

$$\mathrm{Ver} : G^{\mathrm{ab}} \longrightarrow g^{\mathrm{ab}}$$

induced by the restriction $res_{-2} : H^{-2}(G, \mathbb{Z}) \longrightarrow H^{-2}(g, \mathbb{Z})$ *is called the* **Verlagerung** *or* **transfer** *from G to g.*

This canonical homomorphism can also be defined using group theoretic instead of cohomological methods, although this requires some effort and involves quite a bit of formulas. Cf. [16], 14.2.

In addition to restriction, there is another map in the opposite direction

$$cor_q : H^q(g, A) \longrightarrow H^q(G, A),$$

the **corestriction**. As with restriction, corestriction is completely determined once it is given for a single dimension. Nevertheless, before giving the general definition, we explain this map in the two dimensions $q = -1$ and $q = 0$:

In case $q = -1$ we define a homomorphism

$$cor_{-1} : H^{-1}(g, A) \longrightarrow H^{-1}(G, A)$$

by

$$a + I_g A \longmapsto a + I_G A \qquad (a \in {}_{N_g}A \subseteq {}_{N_G}A).$$

In case $q = 0$, we obtain a homomorphism

$$\mathrm{cor}_0 : H^0(g, A) \longrightarrow H^0(G, A).$$

by

$$a + N_g A \longmapsto N_{G/g} a + N_G A \qquad (a \in A^g).$$

Here we let $N_{G/g} a = \sum_{\sigma \in G/g} \sigma a \in A^G$ for $a \in A^g$, where $\sigma \in G/g$ means that σ ranges over a system of left coset representatives of g in G.

The following lemma is the analogon of (4.8) for the corestriction:

(4.11) Lemma. *Let $0 \to A \xrightarrow{i} B \xrightarrow{j} C \to 0$ be an exact sequence of G-modules. Then the following diagram is commutative*

$$
\begin{array}{ccc}
H^{-1}(g, C) & \xrightarrow{\ \delta\ } & H^0(g, A) \\
\downarrow{\scriptstyle \mathrm{cor}_{-1}} & & \downarrow{\scriptstyle \mathrm{cor}_0} \\
H^{-1}(G, C) & \xrightarrow{\ \delta\ } & H^0(G, A)
\end{array}
$$

Proof. Let $c \in {}_{N_g} C$ be a (-1)-cocycle for the class $\bar{c} = c + I_g C \in H^{-1}(g, C)$, thus $c \in {}_{N_G} C$ is a (-1)-cocycle for the class $\mathrm{cor}_{-1} \bar{c} = c + I_G C \in H^{-1}(G, C)$. If we choose $b \in B$ with $jb = c$, we have $j\partial b = \partial c = N_g c = 0$ which implies that there exists a 0-cocycle $a \in A^g$ with $ia = \partial b = N_g b$. By definition $\delta \bar{c} = \bar{a} = a + N_g A$, therefore $\mathrm{cor}_0 \delta \bar{c} = N_{G/g} a + N_G A \in H^0(G, A)$. On the other hand, $\delta \mathrm{cor}_{-1} \bar{c} = \delta(c + I_G C)$. If we choose the same $b \in B$ with $jb = c$ as above, then $\partial b = N_G b = N_{G/g} N_g b = N_{G/g}(ia) = i(N_{G/g} a)$ and we have $\delta(c + I_G C) = N_{G/g} a + N_G A$. Therefore $\mathrm{cor}_0 \delta \bar{c} = N_{G/g} a + N_G A = \delta \, \mathrm{cor}_{-1} \bar{c}$.

Similar to restriction, we define corestriction using an axiomatic approach:

(4.12) Definition. *Let G be a finite group, and let g be a subgroup of G. Then **corestriction** is the uniquely determined family of homomorphisms*

$$\mathrm{cor}_q : H^q(g, A) \longrightarrow H^q(G, A), \quad q \in \mathbb{Z},$$

with the properties:

(i) *If $q = 0$, then*

$$\mathrm{cor}_0 : H^0(g, A) \longrightarrow H^0(G, A), \quad a + N_g A \mapsto N_{G/g} a + N_G A \ (a \in A^g).$$

(ii) *For every exact sequence $0 \longrightarrow A \longrightarrow B \longrightarrow C \longrightarrow 0$ of G-modules and G-homomorphisms, the following diagram is commutative*

$$
\begin{array}{ccc}
H^q(g, C) & \xrightarrow{\ \delta\ } & H^{q+1}(g, A) \\
\downarrow{\scriptstyle \mathrm{cor}_q} & & \downarrow{\scriptstyle \mathrm{cor}_{q+1}} \\
H^q(G, C) & \xrightarrow{\ \delta\ } & H^{q+1}(G, A).
\end{array}
$$

Exactly as for the restrictions, the homomorphisms cor_q arise from the corestriction cor_0 in dimension 0 by dimension shifting:

From (3.15) we have the isomorphisms

$$\delta^q : H^0(G, A^q) \longrightarrow H^q(G, A) \text{ and } \delta^q : H^0(g, A^q) \longrightarrow H^q(g, A),$$

and by (ii) the map cor_q is uniquely determined by the commutative diagram

$$
\begin{array}{ccc}
H^0(g, A^q) & \xrightarrow{\ \delta^q\ } & H^q(g, A) \\
\downarrow{\scriptstyle \mathrm{cor}_0} & & \downarrow{\scriptstyle \mathrm{cor}_q} \\
H^0(G, A^q) & \xrightarrow{\ \delta^q\ } & H^q(G, A).
\end{array}
$$

In particular, because of uniqueness and (4.11) we recover the homomorphism cor_{-1} introduced on p. 38. The fact that (ii) holds is verified in the same way as for restriction using the following diagram, together with (4.11) and (3.6),

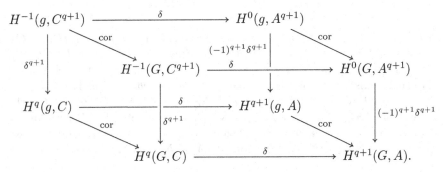

We remark that one can define the corestrictions for negative dimensions very easily by a canonical correspondence between cochains, analogously to the restrictions for positive dimension. However, we will not pursue this further. In view of (4.10) we now want to prove the following theorem:

(4.13) Theorem. *Let $g \subseteq G$ be a subgroup. The homomorphism*

$$\kappa : g^{\mathrm{ab}} \longrightarrow G^{\mathrm{ab}}$$

induced by the corestriction $\mathrm{cor}_{-2} : H^{-2}(g, \mathbb{Z}) \longrightarrow H^{-2}(G, \mathbb{Z})$ *coincides with the canonical homomorphism induced by $\sigma g' \mapsto \sigma G'$.*

This follows, using the proof of (3.19), from the commutative diagram

$$
\begin{array}{ccccc}
H^{-2}(g, \mathbb{Z}) & \xrightarrow{\ \delta\ } & H^{-1}(g, I_g) = I_g/I_g^2 & \xleftarrow[\sim]{\ \log\ } & g^{\mathrm{ab}} \\
\downarrow{\scriptstyle \mathrm{cor}_{-2}} & & \downarrow{\scriptstyle \mathrm{cor}_{-1}} & & \downarrow{\scriptstyle \kappa} \\
H^{-2}(G, \mathbb{Z}) & \xrightarrow{\ \delta\ } & H^{-1}(G, I_G) = I_G/I_G^2 & \xleftarrow[\sim]{\ \log\ } & G^{\mathrm{ab}}.
\end{array}
$$

The following relation between restriction and corestriction is important:

(4.14) Theorem. *Let $g \subseteq G$ be a subgroup. Then the composition*
$$H^q(G, A) \xrightarrow{\text{res}} H^q(g, A) \xrightarrow{\text{cor}} H^q(G, A)$$
is the endomorphism
$$\text{cor} \circ \text{res} = (G : g) \cdot \text{id}.$$

Proof. Consider the case $q = 0$. If $\bar{a} = a + N_G A \in H^0(G, A)$, $a \in A^G$, then $\text{cor}_0 \circ \text{res}_0(\bar{a}) = \text{cor}_0(a + N_g A) = N_{G/g} a + N_G A = (G : g) \cdot a + N_G A = (G : g) \cdot \bar{a}$. The general case follows from this by dimension shifting. In fact, the diagram
$$\begin{array}{ccc} H^0(G, A^q) & \xrightarrow{\text{cor}_0 \circ \text{res}_0} & H^0(G, A^q) \\ {\scriptstyle \delta^q} \downarrow & & \downarrow {\scriptstyle \delta^q} \\ H^q(G, A) & \xrightarrow{\text{cor}_q \circ \text{res}_q} & H^q(G, A) \end{array}$$
commutes, and since the upper horizontal map is $(G : g) \cdot \text{id}$, it follows that the same holds for the lower horizontal map, i.e., $\text{cor}_q \circ \text{res}_q = (G : g) \cdot \text{id}$.

Because the restriction and corestriction maps res and cor commute with the connecting homomorphism δ, they also commute with maps induced by G-homomorphisms:

(4.15) Proposition. *If $f : A \to B$ is a G-homomorphism of the G-modules A, B, and g is a subgroup of G, then the following diagram commutes*
$$\begin{array}{ccc} H^q(G, A) & \xrightarrow{\bar{f}} & H^q(G, B) \\ {\scriptstyle \text{res}} \updownarrow {\scriptstyle \text{cor}} & & {\scriptstyle \text{res}} \updownarrow {\scriptstyle \text{cor}} \\ H^q(g, A) & \xrightarrow{\bar{f}} & H^q(g, B). \end{array}$$

This is clear in case of dimension $q = 0$, and the general case follows easily by dimension shifting. In fact, the homomorphism $f : A \to B$ induces a homomorphism $f : A^q \to B^q$, and in the following diagram

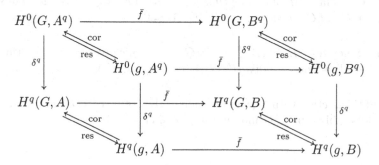

all vertical squares are commutative. Hence the commutativity of the lower diagram follows from that of the upper one.

Since the cohomology groups $H^q(G, A)$ are abelian torsion groups, they are direct sums of their p-**Sylow groups**, i.e., the groups $H^q(G, A)_p$ of all elements in $H^q(G, A)$ of p-power order:

$$H^q(G, A) = \bigoplus_p H^q(G, A)_p.$$

The group $H^q(G, A)_p$ is often called the p-**primary** part of $H^q(G, A)$. For the restriction and corestriction maps on these p-primary parts we have the following:

(4.16) Theorem. *Let A be a G-module, and G_p a p-Sylow subgroup of G. Then the restriction*

$$\mathrm{res} : H^q(G, A)_p \longrightarrow H^q(G_p, A)$$

is injective, and the corestriction

$$\mathrm{cor} : H^q(G_p, A) \longrightarrow H^q(G, A)_p$$

is surjective.

Proof.   Since $\mathrm{cor} \circ \mathrm{res} = (G : G_p) \cdot \mathrm{id}$, and since $(G : G_p)$ and p are relatively prime, the mapping $H^q(G, A)_p \xrightarrow{\mathrm{cor}\, \circ\, \mathrm{res}} H^q(G, A)_p$ is an automorphism. Hence if $x \in H^q(G, A)_p$ and $\mathrm{res}\, x = 0$, it follows immediately from $\mathrm{cor} \circ \mathrm{res}\, x = 0$ that $x = 0$, which shows the injectivity of res on $H^q(G, A)_p$.

On the other hand, $H^q(G_p, A)$ consists of elements whose order is a p-power (cf. (3.16)), so that $\mathrm{cor}\, H^q(G_p, A) \subseteq H^q(G, A)_p$. Since $\mathrm{cor} \circ \mathrm{res}$ is a bijection on $H^q(G, A)_p$, this inclusion is an equality.

We often encounter the problem that we want to show that certain cohomology groups vanish. In many of these cases we will use the following consequence of Theorem (4.16), which reduces this problem to the case of p-groups:

(4.17) Corollary. *If for every prime p the group $H^q(G_p, A) = 0$ for a p-Sylow subgroup G_p of G, then we have $H^q(G, A) = 0$.*

Proof.   Since $\mathrm{res} : H^q(G, A)_p \to H^q(G_p, A)$ is injective, the assumption implies that all p-Sylow groups $H^q(G, A)_p$ are trivial; thus $H^q(G, A) = 0$.

We end this section with a generalization of the notion of a G-induced module; we will use this type of G-modules in global class field theory.

(4.18) Definition. *Let G be a finite group, and let g be a subgroup of G. A G-module A is called G/g-induced, if it has a representation*

$$A = \bigoplus_{\sigma \in G/g} \sigma D,$$

where $D \subseteq A$ is a g-module and σ ranges over a system of left coset representatives of g in G.

For $g = \{1\}$ we obviously recover the G-induced modules from (3.9). As a generalization of the cohomological triviality of G-induced modules, we have the following result, which is often referred to as **Shapiro's Lemma**:

(4.19) Theorem. *Let $A = \bigoplus_{\sigma \in G/g} \sigma D$ be a G/g-induced G-module. Then*

$$H^q(G, A) \cong H^q(g, D);$$

this isomorphism is given by the composition

$$H^q(G, A) \xrightarrow{\text{res}} H^q(g, A) \xrightarrow{\bar\pi} H^q(g, D),$$

where $\bar\pi$ is induced by the natural projection $A \xrightarrow{\pi} D$.

We give a proof using dimension shifting. Let $A = \bigoplus_{i=1}^m \sigma_i D$, where σ_i ranges over a system of left coset representatives of G/g, in particular let $\sigma_1 = 1$. For $q = 0$ we define a map in the opposite direction of the homomorphism

$$A^G/N_G A \xrightarrow{\text{res}} A^g/N_g A \xrightarrow{\bar\pi} D^g/N_g D$$

by $\nu : D^g/N_g D \to A^G/N_G A$, $\nu(d + N_g D) = \sum_{i=1}^m \sigma_i d + N_G A$. It is easy to verify that $(\bar\pi \circ \text{res}) \circ \nu = \text{id}$ and $\nu \circ (\bar\pi \circ \text{res}) = \text{id}$. Therefore $\bar\pi \circ \text{res}$ is bijective.

In case of arbitrary dimension q we now set

$$A^q = J_G \otimes \cdots \otimes J_G \otimes A \qquad\qquad A^q = I_G \otimes \cdots \otimes I_G \otimes A$$
$$D_*^q = J_G \otimes \cdots \otimes J_G \otimes D \quad \text{resp.} \quad D_*^q = I_G \otimes \cdots \otimes I_G \otimes D$$
$$D^q = J_g \otimes \cdots \otimes J_g \otimes D \qquad\qquad D^q = I_g \otimes \cdots \otimes I_g \otimes D$$

depending on whether $q \geq 0$ or $q \leq 0$. Because $A = \bigoplus_{i=1}^m \sigma_i D$ we have

$$J_G = J_g \oplus K_1 \qquad \text{resp.} \qquad I_G = I_g \oplus K_{-1}$$

with the g-induced modules

$$K_1 = \bigoplus_{\tau \in g} \tau \Big(\sum_{i=2}^m \mathbb{Z} \cdot \bar\sigma_i^{-1} \Big) \quad \text{and} \quad K_{-1} = \bigoplus_{\tau \in g} \tau \Big(\sum_{i=2}^m \mathbb{Z} \cdot (\sigma_i^{-1} - 1) \Big).$$

With (1.5) and (3.10) we obtain for all q the canonical g-module decomposition

$$D_*^q = D^q \oplus C^q$$

for some g-induced g-module C^q. Using (3.15), we then obtain the diagram

$$\begin{array}{ccccccc}
H^0(G, A^q) & \xrightarrow{\text{res}} & H^0(g, A^q) & \xrightarrow{\bar{\pi}_*} & H^0(g, D_*^q) & \xrightarrow{\bar{\rho}} & H^0(g, D^q) \\
\Big\downarrow{\delta^q} & & \Big\downarrow{\delta^q} & & & & \Big\downarrow{\delta^q} \\
H^q(G, A) & \xrightarrow{\text{res}} & H^q(g, A) & & \xrightarrow{\qquad \bar{\pi} \qquad} & & H^q(g, D) \ ,
\end{array}$$

in which the map $\bar{\pi}_* \circ \text{res}$ in the upper row in dimension 0 is bijective, and the following map $\bar{\rho}$ is bijective because of (3.7) and (3.13). Since the composite $A^q \xrightarrow{\pi_*} D_*^q \xrightarrow{\rho} D^q$ is induced by the projection $A \xrightarrow{\pi} D$, we see that this diagram commutes. Thus the bijectivity of the upper map $\bar{\rho} \circ \bar{\pi}_* \circ \text{res}$ implies the bijectivity of the lower map $\bar{\pi} \circ \text{res}$.

§ 5. The Cup Product

In the previous section we have seen that the restriction and corestriction maps are given by canonical data in dimension $q = 0$, and induce corresponding maps on cohomology in all dimensions. The same principle applies to the **cup product**, which in dimension 0 is just the **tensor product**.

Let A and B be G-modules. Then $A \otimes B$ is a G-module, and the map $(a, b) \mapsto a \otimes b$ induces a canonical bilinear mapping

$$A^G \times B^G \longrightarrow (A \otimes B)^G,$$

which maps $N_G A \times N_G B$ to $N_G(A \otimes B)$. Hence it induces a bilinear mapping

$$H^0(G, A) \times H^0(G, B) \longrightarrow H^0(G, A \otimes B) \text{ by } (\bar{a}, \bar{b}) \mapsto \overline{a \otimes b} \ ^{12)}.$$

We call the element $\overline{a \otimes b} \in H^0(G, A \otimes B)$ the **cup product** of $\bar{a} \in H^0(G, A)$ and $\bar{b} \in H^0(G, B)$, and denote it by

$$\bar{a} \cup \bar{b} = \overline{a \otimes b}.$$

This cup product in dimension 0 extends to arbitrary dimensions:

(5.1) Definition. *There exists a uniquely determined family of bilinear mappings, the* **cup product**

$$\cup : H^p(G, A) \times H^q(G, B) \longrightarrow H^{p+q}(G, A \otimes B), \ p, q \in \mathbb{Z},$$

with the following properties:

---

[12] As usual, we denote by $\bar{a}$ the cohomology class $\bar{a} = a + N_G A$ of the element $a \in A^G$; similar for $\bar{b}$. Likewise, $\overline{a \otimes b}$ stands for the cohomology class $\overline{a \otimes b} = a \otimes b + N_G(A \otimes B)$ of $a \otimes b \in (A \otimes B)^G$.

(i) For $p = q = 0$ the cup product is given by
$$(\bar{a}, \bar{b}) \longmapsto \bar{a} \cup \bar{b} = \overline{a \otimes b}, \quad \bar{a} \in H^0(G, A), \ \bar{b} \in H^0(G, B).$$

(ii) If the sequences of G-modules
$$0 \longrightarrow A \longrightarrow A' \longrightarrow A'' \longrightarrow 0$$
$$0 \longrightarrow A \otimes B \longrightarrow A' \otimes B \longrightarrow A'' \otimes B \longrightarrow 0$$

are both exact, then the following diagram commutes

$$
\begin{array}{ccc}
H^p(G, A'') \times H^q(G, B) & \overset{\cup}{\longrightarrow} & H^{p+q}(G, A'' \otimes B) \\
{\scriptstyle \delta} \downarrow \quad\quad {\scriptstyle 1} \downarrow & & \downarrow {\scriptstyle \delta} \\
H^{p+1}(G, A) \times H^q(G, B) & \overset{\cup}{\longrightarrow} & H^{p+q+1}(G, A \otimes B)
\end{array}
$$

so that $\delta(\bar{a}'' \cup \bar{b}) = \delta\bar{a}'' \cup \bar{b}, \quad \bar{a}'' \in H^p(G, A''), \bar{b} \in H^q(G, B).$

(iii) If the sequences of G-modules
$$0 \longrightarrow B \longrightarrow B' \longrightarrow B'' \longrightarrow 0,$$
$$0 \longrightarrow A \otimes B \longrightarrow A \otimes B' \longrightarrow A \otimes B'' \longrightarrow 0$$

are both exact, then the following diagram commutes

$$
\begin{array}{ccc}
H^p(G, A) \times H^q(G, B'') & \overset{\cup}{\longrightarrow} & H^{p+q}(G, A \otimes B'') \\
{\scriptstyle 1} \downarrow \quad\quad {\scriptstyle \delta} \downarrow & & \downarrow {\scriptstyle (-1)^p \delta} \\
H^p(G, A) \times H^{q+1}(G, B) & \overset{\cup}{\longrightarrow} & H^{p+q+1}(G, A \otimes B)
\end{array}
$$

i.e., we have $\delta(\bar{a} \cup \bar{b}'') = (-1)^p(\bar{a} \cup \delta\bar{b}''), \quad \bar{a} \in H^p(G, A), \bar{b}'' \in H^q(G, B'').$

The factor $(-1)^p$ in the last diagram is necessary and results from the anticommutativity of the connecting homomorphism δ, see below. One cannot define a reasonable cup product omitting this factor.

As with the general restriction maps, we obtain the general cup product from the case $p = 0$, $q = 0$ by dimension shifting[13].

We recall that we identify the G-modules $A \otimes B$ and $B \otimes A$, as well as the G-modules $(A \otimes B) \otimes C$ and $A \otimes (B \otimes C)$ (cf. §1, p. 7). This automatically leads to a corresponding identification of the cohomology groups of these G-modules. In particular, we can write (cf. §3, p. 29):

---

[13] The readers who are mainly interested in applications of the cohomological calculus won't lose much by omitting the details of this shifting process. They will be satisfied with the functorial behavior of the cup product and with its explicit description in small dimensions (cf. (5.2), (5.6), (5.7) and (5.8)).

$$A^p \otimes B = J_G \otimes \cdots \otimes J_G \otimes A \otimes B = (A \otimes B)^p \quad \text{and}$$

$$A \otimes B^q = A \otimes J_G \otimes \cdots \otimes J_G \otimes B = J_G \otimes \cdots \otimes J_G \otimes A \otimes B = (A \otimes B)^q$$

for $p, q \geq 0$, and analogously for $p, q \leq 0$ with I_G in place of J_G. We will use this freely below.

Because of Proposition (3.15) we may start with the case $q = 0$, $p = 0$ and determine the cup product by the following commutative diagram:

$$\begin{array}{ccc}
H^0(G, A^p) \times H^0(G, B^q) & \xrightarrow{\cup} & H^0(G, (A \otimes B^q)^p) = H^0(G, A^p \otimes B^q) \\
{\scriptstyle \delta^p}\downarrow \qquad\qquad \downarrow{\scriptstyle 1} & & \downarrow{\scriptstyle \delta^p} \\
H^p(G, A) \times H^0(G, B^q) & \xrightarrow{\cup} & H^p(G, (A \otimes B)^q) = H^p(G, A \otimes B^q) \\
{\scriptstyle 1}\downarrow \qquad\qquad \downarrow{\scriptstyle \delta^q} & & \downarrow{\scriptstyle (-1)^{p \cdot q}\delta^q} \\
H^p(G, A) \times H^q(G, B) & \xrightarrow{\cup} & H^{p+q}(G, A \otimes B)
\end{array}$$

(∗)

It follows immediately from the conditions (i), (ii) and (iii) that the cup product is unique. We use this fact to give an explicit description of the cup product in terms of cocycles in the special cases $(p, q) = (0, q)$ and $(p, 0)$:

(5.2) Proposition. If we denote by a_p (resp. b_q) p-cocycles (resp. q-cocycles) of A (resp. B), and by $\bar{a}_p$ (resp. $\bar{b}_q$) their cohomology classes, then

$$\bar{a}_0 \cup \bar{b}_q = \overline{a_0 \otimes b_q} \quad \text{and} \quad \bar{a}_p \cup \bar{b}_0 = \overline{a_p \otimes b_0} \quad {}^{14)}.$$

For the proof note that the products $\bar{a}_0 \cup \bar{b}_q$ and $\bar{a}_p \cup \bar{b}_0$ defined here satisfy the conditions (i), (ii) and (iii) for $(0, q)$ and $(p, 0)$ respectively. This can be seen directly from the behaviour of the cocycles under the corresponding maps. Now if we consider the lower part of the diagram (∗) for $p = 0$, resp. the upper part for $q = 0$, then we see that the product defined by the commutative diagram (∗) must coincide with the one defined by (5.2).

Thus everything boils down to showing that the product maps defined by (∗)

$$H^p(G, A) \times H^q(G, B) \xrightarrow{\cup} H^{p+q}(G, A \otimes B)$$

satisfy the conditions (ii) and (iii). To this end, consider the exact sequences

$$0 \longrightarrow A \longrightarrow A' \longrightarrow A'' \longrightarrow 0,$$

$$0 \longrightarrow A \otimes B \longrightarrow A' \otimes B \longrightarrow A'' \otimes B \longrightarrow 0$$

[14)] Note that if $b_q(\sigma_1, \ldots, \sigma_q) \in B$ is a q-cocycle, then $a_0 \otimes b_q(\sigma_1, \ldots, \sigma_q) \in A \otimes B$ ($a_0 \in A^G$) is also a q-cocycle.

and
$$0 \longrightarrow B \longrightarrow B' \longrightarrow B'' \longrightarrow 0,$$
$$0 \longrightarrow A \otimes B \longrightarrow A \otimes B' \longrightarrow A \otimes B'' \longrightarrow 0.$$

From these we get by (1.9) and (1.2) the exact sequences
$$0 \longrightarrow A^q \longrightarrow A'^q \longrightarrow A''^q \longrightarrow 0,$$
$$0 \longrightarrow (A \otimes B)^q \longrightarrow (A' \otimes B)^q \longrightarrow (A'' \otimes B)^q \longrightarrow 0$$

and
$$0 \longrightarrow B^p \longrightarrow B'^p \longrightarrow B''^p \longrightarrow 0,$$
$$0 \longrightarrow (A \otimes B)^p \longrightarrow (A \otimes B')^p \longrightarrow (A \otimes B'')^p \longrightarrow 0,$$

and we have the diagrams

and

Here the left sides in both diagrams commute for trivial reasons. The right sides are composed from q (resp. p) squares as in (3.6), thus they commute as well. The front and back sides commute by definition ($*$) of the cup product, and the upper squares commute because of (5.2) and the remarks following it. Since the vertical maps are bijective, the commutativity of the upper squares implies the commutativity of the lower squares. This completes the proof.

The axiomatic definition of the cup product in (5.1) does not give us an explicit description of it; i.e., given two cohomology classes in terms of cocyles, we are for now not in a position to decide which cocyle represents their cup product in general. Only for the cases $(p, q) = (0, q)$ and $(p, 0)$ we have such a description by (5.2). The attempt to give an explicit description of the cup product for general p, q (in particular for $p < 0$ and $q < 0$) leads, however, to major computational problems. Thus we find ourselves in a situation which is similar to that of the restriction map, which admits a very simple description in dimensions $q \geq 0$, but not for negative dimensions. Nevertheless in both cases we will need explicit computations only in low dimensions; given these, one can manage knowing the functorial properties of these maps.

Before giving explicit formulas for small dimensions, we want to convince ourselves that the cup product is compatible with the usual cohomological maps defined above.

(5.3) Proposition. Let $f : A \to A'$ and $g : B \to B'$ be two G-homomorphisms, and let $f \otimes g : A \otimes B \to A' \otimes B'$ be the G-homomorphism induced by f and g. If $\bar{a} \in H^p(G, A)$ and $\bar{b} \in H^q(G, B)$, then

$$\bar{f}(\bar{a}) \cup \bar{g}(\bar{b}) = \overline{f \otimes g}\,(\bar{a} \cup \bar{b}) \in H^{p+q}(G, A' \otimes B').$$

This is completely trivial for $p = q = 0$, and follows in general from a simple dimension shifting argument. We have demonstrated this technique already frequently enough to leave the details to the reader.

(5.4) Proposition. Let A, B be G-modules, and let g be a subgroup of G. If $\bar{a} \in H^p(G, A)$ and $\bar{b} \in H^q(G, B)$, then

$$\mathrm{res}(\bar{a} \cup \bar{b}) = \mathrm{res}\,\bar{a} \cup \mathrm{res}\,\bar{b} \in H^{p+q}(g, A \otimes B),$$

and

$$\mathrm{cor}(\mathrm{res}\,\bar{a} \cup \bar{b}) = \bar{a} \cup \mathrm{cor}\,\bar{b} \in H^{p+q}(G, A \otimes B).$$

This follows again from the case $p = q = 0$ by dimension shifting. In case $p = q = 0$ the first formula is immediate. For the second, let $a \in A^G$ and $b \in B^g$ be 0-cocycles representing $\bar{a}$ and $\bar{b}$ respectively. By definition (4.12) of the corestriction in dimension 0, we have

$$\text{cor}(\text{res}\,\overline{a} \cup \overline{b}) = \text{cor}(a \otimes b + N_g(A \otimes B))$$

$$= \sum_{\sigma \in G/g} \sigma(a \otimes b) + N_G(A \otimes B)$$

$$= \sum_{\sigma \in G/g} a \otimes \sigma b + N_G(A \otimes B)$$

$$= a \otimes (\sum_{\sigma \in G/g} \sigma b) + N_G(A \otimes B)$$

$$= \overline{a} \cup \text{cor}\,\overline{b}.$$

We show that the cup product is anticommutative and associative:

(5.5) Theorem. *Let* $\overline{a} \in H^p(G, A)$, $\overline{b} \in H^q(G, B)$, *and* $\overline{c} \in H^r(G, C)$. *Then*
$$\overline{a} \cup \overline{b} = (-1)^{p \cdot q}(\overline{b} \cup \overline{a}) \in H^{p+q}(G, B \otimes A),$$

and
$$(\overline{a} \cup \overline{b}) \cup \overline{c} = \overline{a} \cup (\overline{b} \cup \overline{c}) \in H^{p+q+r}(G, A \otimes (B \otimes C))$$

under the canonical isomorphisms $H^{p+q}(G, A \otimes B) \cong H^{p+q}(G, B \otimes A)$ *and* $H^{p+q+r}(G, (A \otimes B) \otimes C) = H^{p+q+r}(G, A \otimes (B \otimes C))$. [15]

Again, this is trivial for $p = q = 0$, and follows in general by dimension shifting.

We now want to compute some explicit formulas for the cup product. For this we denote by a_p (resp. b_q) p-cocycles of A (resp. q-cocycles of B), and write $\overline{a}_p$ (resp. $\overline{b}_q$) for their cohomology classes in $H^p(G, A)$ (resp. $H^q(G, B)$).

(5.6) Lemma. *We have* $\overline{a}_1 \cup \overline{b}_{-1} = \overline{x}_0 \in H^0(G, A \otimes B)$ *with*
$$x_0 = \sum_{\tau \in G} a_1(\tau) \otimes \tau b_{-1}.$$

Proof. By (3.14) we have the G-induced G-module $A' = \mathbb{Z}[G] \otimes A$ and the exact sequences

$$0 \longrightarrow A \longrightarrow A' \longrightarrow A'' \longrightarrow 0,$$
$$0 \longrightarrow A \otimes B \longrightarrow A' \otimes B \longrightarrow A'' \otimes B \longrightarrow 0.$$

We think of A embedded in A' and $A \otimes B$ embedded in $A' \otimes B$; to simplify notation we do not explicitly write out these homomorphisms. Because of the vanishing $H^1(G, A') = 0$, there is a 0-cochain $a_0' \in A'$ with $a_1 = \partial a_0'$, so that

---

[15] More precisely, one should say that $(-1)^{p \cdot q}(\overline{b} \cup \overline{a})$ is the image of $\overline{a} \cup \overline{b}$ under the canonical isomorphism $H^{p+q}(G, A \otimes B) \cong H^{p+q}(G, B \otimes A)$ induced by $A \otimes B \cong B \otimes A$, and similarly for the second formula. Cf. §1, p. 7.

$(*)$ $\qquad\qquad a_1(\tau) = \tau a_0' - a_0' \quad \text{for all} \quad \tau \in G.$

Let $a_0'' \in A''^G$ be the image of a_0' in A''. By definition of the connecting homomorphism δ, we have $\bar{a}_1 = \delta(a_0'')$, and we obtain

$$\bar{a} \cup \bar{b}_{-1} = \delta(\overline{a_0''}) \cup \bar{b}_{-1} \overset{(5.1)}{=} \delta(\overline{a_0'' \cup \bar{b}_{-1}}) \overset{(5.2)}{=} \delta(\overline{a_0'' \otimes b_{-1}}) = \overline{\partial(a_0' \otimes b_{-1})}$$

$$= \overline{N_G(a_0' \otimes b_{-1})} = \overline{\sum_{\tau \in G} \tau a_0' \otimes \tau b_{-1}} \overset{(*)}{=} \overline{\sum_{\tau \in G} (a_1(\tau) + a_0') \otimes \tau b_{-1}}$$

$$= \overline{\sum_{\tau \in G} (a_1(\tau) \otimes \tau b_{-1}) + a_0' \otimes N_G b_{-1}} = \overline{\sum_{\tau \in G} (a_1(\tau) \otimes \tau b_{-1})}$$

because $N_G b_{-1} = 0$.

In the following we restrict to the case $B = \mathbb{Z}$ and identify $A \otimes \mathbb{Z}$ with A via $a \otimes n \mapsto a \cdot n$. Recall that from (3.19) we have the canonical isomorphism

$$H^{-2}(G, \mathbb{Z}) \cong G^{\mathrm{ab}}.$$

If $\sigma \in G$, let $\bar{\sigma}$ be the element in $H^{-2}(G, \mathbb{Z})$ corresponding to $\sigma \cdot G' \in G^{\mathrm{ab}}$.

(5.7) Lemma. $\bar{a}_1 \cup \bar{\sigma} = \overline{a_1(\sigma)} \in H^{-1}(G, A).$

Proof. From the exact sequence

$$0 \longrightarrow A \otimes I_G \longrightarrow A \otimes \mathbb{Z}[G] \longrightarrow A \longrightarrow 0$$

we obtain the isomorphism $H^{-1}(G, A) \overset{\delta}{\longrightarrow} H^0(G, A \otimes I_G)$. Thus it suffices to show $\delta(\bar{a}_1 \cup \bar{\sigma}) = \delta(\overline{a_1(\sigma)})$. Using the definition of δ, we now compute

$$\delta(\overline{a_1(\sigma)}) = \overline{x}_0 \quad \text{with} \quad x_0 = \sum_{\tau \in G} \tau a_1(\sigma) \otimes \tau.$$

On the other hand, the proof of (3.19) shows that under the isomorphism $H^{-2}(G, \mathbb{Z}) \overset{\delta}{\to} H^{-1}(G, I_G)$ the element $\bar{\sigma}$ goes to $\delta \bar{\sigma} = \overline{\sigma - 1}$, hence we have

$$\delta(\bar{a}_1 \cup \bar{\sigma}) \overset{(5.1)}{=} -(\bar{a}_1 \cup \delta(\bar{\sigma})) = -\bar{a}_1 \cup \overline{(\sigma - 1)} = \bar{y}_0.$$

For the cocyle y_0 we obtain from (5.6)

$$y_0 = -\sum_{\tau \in G} a_1(\tau) \otimes \tau(\sigma - 1) = \sum_{\tau \in G} a_1(\tau) \otimes \tau - \sum_{\tau \in G} a_1(\tau) \otimes \tau\sigma.$$

The 1-cocycle $a_1(\tau)$ satisfies $a_1(\tau) = a_1(\tau\sigma) - \tau a_1(\sigma)$. Substituting this into the last sum, we find

$$y_0 = \sum_{\tau \in G} \tau a_1(\sigma) \otimes \tau\sigma.$$

Therefore $y_0 - x_0 = \sum_{\tau \in G} \tau a_1(\sigma) \otimes \tau(\sigma - 1) = N_G(a_1(\sigma) \otimes (\sigma - 1))$, which shows that $\bar{x}_0 = \bar{y}_0$.

The following formula (5.8) is of particular interest for us. Note that if we take an element $\bar{a}_2$ in the group $H^2(G, A)$, it provides us with the homomorphism

$$\bar{a}_2\cup : H^{-2}(G, \mathbb{Z}) \longrightarrow H^0(G, A),$$

which maps each $\bar{\sigma} \in H^{-2}(G, \mathbb{Z})$ to the cup product $\bar{a}_2 \cup \bar{\sigma} \in H^0(G, A)$; we thus get a canonical mapping from the abelianization G^{ab} to the norm residue group $A^G/N_G A$. In class field theory we will consider a special G-module A for which this homomorphism will be shown to be bijective; in fact, the resulting canonical isomorphism $G^{ab} \cong A^G/N_G A$ is the main theorem of class field theory. For this the following proposition will be important:

(5.8) Proposition. We have $\bar{a}_2 \cup \bar{\sigma} = \overline{\sum_{\tau \in G} a_2(\tau, \sigma)} \in H^0(G, A)$.

Proof. We consider again the G-module $A' = \mathbb{Z}[G] \otimes A$ and the exact sequence $0 \to A \to A' \to A'' \to 0$ ($A'' = J_G \otimes A$). Since $H^2(G, A') = 0$ there is a 1-cochain $a_1' \in A_1'$ with $a_2 = \partial a_1'$ i.e.,

$$(*) \qquad\qquad a_2(\tau, \sigma) = \tau a_1'(\sigma) - a_1'(\tau \cdot \sigma) + a_1'(\tau).$$

The image a_1'' of a_1' is a 1-cocycle of A'' such that $\bar{a}_2 = \delta(\overline{a_1''})$. Therefore

$$\bar{a}_2 \cup \bar{\sigma} = \delta(\overline{a_1''}) \cup \bar{\sigma} \overset{(5.1)}{=} \delta(\overline{a_1'' \cup \bar{\sigma}}) \overset{(5.7)}{=} \delta(\overline{a_1''(\sigma)}) = \overline{\partial(a_1'(\sigma))} = \overline{\sum_{\tau \in G} \tau a_1'(\sigma)}$$

$$\overset{(*)}{=} \overline{\sum_{\tau \in G} a_2(\tau, \sigma)} + \overline{\sum_{\tau \in G} a_1'(\tau \cdot \sigma)} - \overline{\sum_{\tau \in G} a_1'(\tau)} = \overline{\sum_{\tau \in G} a_2(\tau, \sigma)}.$$

§ 6. Cohomology of Cyclic Groups

So far we have introduced the basic cohomological maps and have studied their functorial and compatibility properties. Now we will begin to prove the central theorems of cohomology theory. We start with G-modules A, where G is a cyclic group; the cohomology of these G-modules is particularly simple.

Let G be a cyclic group of order n with generator σ. Then we have for the group ring

$$\mathbb{Z}[G] = \bigoplus_{i=0}^{n-1} \mathbb{Z}\sigma^i, \quad N_G = 1 + \sigma + \cdots + \sigma^{n-1},$$

and because $\sigma^k - 1 = (\sigma - 1)(\sigma^{k-1} + \ldots + \sigma + 1)$ ($k \geq 1$), the augmentation ideal I_G is the principal ideal of $\mathbb{Z}[G]$ generated by $\sigma - 1$, i.e., $I_G = \mathbb{Z}[G] \cdot (\sigma - 1)$.

(6.1) Theorem. Let G be a cyclic group and let A be a G-module. Then

$$H^q(G, A) \cong H^{q+2}(G, A) \text{ for all } q \in \mathbb{Z}.$$

Proof. It suffices to specify an isomorphism $H^{-1}(G, A) \cong H^1(G, A)$. Given this, the general case follows from this by dimension shifting (cf. (3.15)), since

$$H^q(G, A) \cong H^{-1}(G, A^{q+1}) \cong H^1(G, A^{q+1}) \cong H^{q+2}(G, A).$$

The group Z_1 of 1-cocycles consists of all the crossed homomorphisms of G in A; thus if $x \in Z_1$, then

$$x(\sigma^k) = \sigma x(\sigma^{k-1}) + x(\sigma) = \sigma^2 x(\sigma^{k-2}) + \sigma x(\sigma) + x(\sigma)$$
$$= \sum_{i=0}^{k-1} \sigma^i x(\sigma) \quad (k \geq 1), \text{ and}$$
$$x(1) = 0 \text{ because } x(1) = x(1) + x(1).$$

It follows that $N_G x(\sigma) = \sum_{i=0}^{n-1} \sigma^i x(\sigma) = x(\sigma^n) = x(1) = 0$, i.e., $x(\sigma) \in {}_{N_G}A$. Conversely, it is easy to see that if $a \in {}_{N_G}A = Z_{-1}$ is a (-1)-cocyle, then

$$x(\sigma) = a \quad \text{and} \quad x(\sigma^k) = \sum_{i=0}^{k-1} \sigma^i a$$

defines a 1-cocycle. Therefore the map

$$x \longmapsto x(\sigma)$$

is an isomorphism from Z_1 to $Z_{-1} = {}_{N_G}A$. Under this isomorphism the group R_1 of 1-coboundaries is mapped to the group R_{-1} of (-1)-coboundaries:

$$x \in R_1 \Longleftrightarrow x(\sigma^k) = \sigma^k a - a \text{ with fixed } a \in A$$
$$\Longleftrightarrow x(\sigma) = \sigma a - a$$
$$\Longleftrightarrow x(\sigma) \in I_G A = R_{-1}.$$

Thus in the case of a cyclic group G we always have isomorphisms

$$H^{2q}(G, A) \cong H^0(G, A) \quad \text{and} \quad H^{2q+1}(G, A) \cong H^1(G, A).$$

If

$$0 \longrightarrow A \longrightarrow B \longrightarrow C \longrightarrow 0$$

is an exact sequence of G-modules, we write the corresponding long exact cohomology sequence in the form of an exact hexagon:

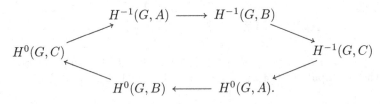

For exactness at the term $H^{-1}(G, A)$, note that the isomorphism $H^1(G, A) \cong H^{-1}(G, A)$ from the proof of (6.1) fits into the commutative diagram

$$H^{-1}(G,A) \longrightarrow H^{-1}(G,B)$$

$$\Big\downarrow\wr \qquad\qquad \Big\downarrow\wr$$

$$H^1(G,A) \longrightarrow H^1(G,B),$$

so that the kernel of the map $H^1(G,A) \to H^1(G,B)$ corresponds to the kernel of the map $H^{-1}(G,A) \to H^{-1}(G,B)$.

For many index and order considerations the notion of a **Herbrand quotient** is very useful; in particular, it can be used to simplify the computations of indices in abelian groups. Although it is of particular interest for G-modules when G is a cyclic group, we want to introduce it in its most general form.

(6.2) Definition. *Let A be an abelian group, and let f, g endomorphisms of A such that $f \circ g = g \circ f = 0$, so that we have inclusions $\operatorname{im} g \subseteq \ker f$ and $\operatorname{im} f \subseteq \ker g$. Then the **Herbrand quotient** is defined as*

$$q_{f,g}(A) = \frac{(\ker f : \operatorname{im} g)}{(\ker g : \operatorname{im} f)}$$

provided both indices are finite.

We are mainly interested in the following special case:

Let A be a G-module with G cyclic of order n. Consider the endomorphisms

$$f = D = \sigma - 1 \quad \text{and} \quad g = N = 1 + \sigma + \cdots + \sigma^{n-1},$$

where σ is a generator of G. Obviously we have

$$D \circ N = N \circ D = 0,$$

and

$$\ker D = A^G, \quad \operatorname{im} N = N_G A; \quad \ker N = {}_{N_G}A, \quad \operatorname{im} D = I_G A.$$

Hence if both cohomology groups $H^0(G,A)$ and $H^{-1}(G,A)$ are finite, then

$$q_{D,N}(A) = \frac{|H^0(G,A)|}{|H^{-1}(G,A)|} = \frac{|H^2(G,A)|}{|H^1(G,A)|}.$$

If this holds, we call A a **Herbrand module**. For these special Herbrand quotients $q_{D,N}(A)$ we want to use the following notation:

(6.3) Definition. *Let G be a cyclic group and A a G-module. Then*

$$h(A) = \frac{|H^0(G,A)|}{|H^{-1}(G,A)|} = \frac{|H^2(G,A)|}{|H^1(G,A)|}.$$

These special Herbrand quotients $h(-)$ are **multiplicative**:

(6.4) Theorem. *Let G be a cyclic group and*

$$0 \longrightarrow A \longrightarrow B \longrightarrow C \longrightarrow 0$$

an exact sequence of G-modules. Then

$$h(B) = h(A) \cdot h(C)$$

in the sense that if two of these quotients are defined, then so is the third, and equality holds.

Proof.   Consider the long exact cohomology sequence, written as the hexagon

$$\begin{array}{ccc} & H^{-1}(G,A) \xrightarrow{\ f_1\ } H^{-1}(G,B) & \\ f_6 \nearrow & & \searrow f_2 \\ H^0(G,C) & & H^{-1}(G,C) \\ f_5 \nwarrow & & \swarrow f_3 \\ & H^0(G,B) \xleftarrow[\ f_4\ ]{} H^0(G,A) & \end{array}$$

If we write F_i for the order of the image of f_i, then

$$|H^{-1}(G,A)| = F_6 \cdot F_1, \quad |H^{-1}(G,B)| = F_1 \cdot F_2, \quad |H^{-1}(G,C)| = F_2 \cdot F_3,$$
$$|H^0(G,A)| = F_3 \cdot F_4, \quad |H^0(G,B)| = F_4 \cdot F_5, \quad |H^0(G,C)| = F_5 \cdot F_6,$$

and therefore

$$(*) \qquad \begin{aligned} |H^{-1}(G,A)| \cdot |H^{-1}(G,C)| \cdot |H^0(G,B)| \\ = |H^{-1}(G,B)| \cdot |H^0(G,A)| \cdot |H^0(G,C)|. \end{aligned}$$

Hence whenever two of the three quotients $h(A)$, $h(B)$, $h(C)$ are defined, then so is the third, and the identity $(*)$ implies the formula $h(B) = h(A) \cdot h(C)$.

Another special case of a Herbrand quotients occurs when A is an abelian group and f and g are the endomorphisms $f = 0$ and $g = n$ (n a positive integer), i.e., g is the map 'multiplication by n' $a \mapsto n \cdot a \in A$. Then

$$q_{0,n}(A) = \frac{(A : nA)}{|{}_nA|} \qquad ({}_nA = \{a \in A \mid n \cdot a = 0\}).$$

In fact, this is just a special case of what we considered above:

(6.5) Proposition. *If the cyclic group G of order n acts trivially on A, then*

$$h(A) = q_{0,n}(A).$$

In particular, the Herbrand quotients $q_{0,n}$ are multiplicative[16]:

---

[16] We remark that under certain assumptions one can show multiplicativity for general Herbrand quotients $q_{f,g}$

(6.6) Proposition. *If* $0 \to A \to B \to C \to 0$ *is an exact sequence of abelian groups, then*
$$q_{0,n}(B) = q_{0,n}(A) \cdot q_{0,n}(C);$$
this again in the sense that the existence of two of these quotients implies the existence of the third.

(6.7) Proposition. *If* A *is a finite group, then we always have*
$$q_{f,g}(A) = 1.$$

Proof. Because of the isomorphisms $\operatorname{im} f \cong A/\ker f$ and $\operatorname{im} g \cong A/\ker g$,
$$|A| = |\ker f| \cdot |\operatorname{im} f| = |\ker g| \cdot |\operatorname{im} g|,$$
which implies the claim.

In particular, a finite G-module A has Herbrand quotient $h(A) = 1$. This remark, together with the multiplicativity shown in (6.4), implies the following:

> If A is a submodule of finite index in the G-module B, then
> $$h(B) = h(A).$$

It is in fact this statement that is most useful in applications of the Herbrand quotient. If the direct computation of the order of the cohomology groups of a G-module B is not possible, the above fact allows us to consider without loss an appropriate submodule A, provided it has finite index. This type of consideration historically motivated the definition of the Herbrand quotient.

In the following we will show how to determine h in case of a cyclic group G of prime order p from the Herbrand quotients $q_{0,p}$. For this we need:

(6.8) Lemma. *Let* g *and* f *be two endomorphisms of an abelian group* A *such that* $f \circ g = g \circ f$. *Then*
$$q_{0,gf}(A) = q_{0,g}(A) \cdot q_{0,f}(A),$$
where again all three quotients are defined whenever any two of them are.

Proof. Consider the commutative diagram with exact rows

$$
\begin{array}{ccccccccc}
0 & \longrightarrow & g(A) \cap \ker f & \longrightarrow & g(A) & \xrightarrow{\ f\ } & fg(A) & \longrightarrow & 0 \\
 & & \downarrow & & \downarrow & & \downarrow & & \\
0 & \longrightarrow & \ker f & \longrightarrow & A & \xrightarrow{\ f\ } & f(A) & \longrightarrow & 0
\end{array}
$$

We obtain the exact sequence

$$0 \longrightarrow \ker f/g(A) \cap \ker f \longrightarrow A/g(A) \longrightarrow f(A)/fg(A) \longrightarrow 0,$$

so that
$$\frac{(A : fg(A))}{(A : f(A))} = \frac{(A : g(A)) \cdot |g(A) \cap \ker f|}{|\ker f|}.$$

If we observe that
$$\ker fg/\ker g = g^{-1}(g(A) \cap \ker f)/g^{-1}(0) \cong g(A) \cap \ker f,$$

we in fact get
$$\frac{(A : gf(A))}{|\ker gf|} = \frac{(A : g(A))}{|\ker g|} \cdot \frac{(A : f(A))}{|\ker f|}.$$

It is easy to verify that all three quotients are defined, if two of them are.

Now we prove the important

(6.9) Theorem. *Let G be a cyclic group of prime order p and A a G-module. If $q_{0,p}(A)$ is defined, then $q_{0,p}(A^G)$ and $h(A)$ are also defined, and*
$$h(A)^{p-1} = q_{0,p}(A^G)^p/q_{0,p}(A).$$

Proof. Let σ be a generator of G and $D = \sigma - 1$. Consider the exact sequence
$$0 \longrightarrow A^G \longrightarrow A \xrightarrow{D} I_G A \longrightarrow 0 .$$

From the fact that $I_G A$ is a subgroup as well as a factor group of A, we conclude immediately that if $q_{0,p}(A)$ is defined, then $q_{0,p}(I_G A)$ is also defined. Hence as a consequence of (6.6), $q_{0,p}(A^G)$ is also defined, and we have

(*) $q_{0,p}(A) = q_{0,p}(A^G) \cdot q_{0,p}(I_G A).$

Since G acts trivially on A^G, it follows from (6.5) that $q_{0,p}(A^G) = h(A^G)$.

To determine the quotient $q_{0,p}(I_G A)$ we use the following interesting trick. Since the ideal $\mathbb{Z} \cdot N_G = \mathbb{Z}(\sum_{i=0}^{p-1} \sigma^i)$ annihilates the module $I_G A$, we can consider $I_G A$ as a $\mathbb{Z}[G]/\mathbb{Z} \cdot N_G$-module. Now the ring $\mathbb{Z}[G]/\mathbb{Z} \cdot N_G$ is isomorphic to the ring $\mathbb{Z}[X]/(1+X+\ldots+X^{p-1})$ with an indeterminate X. But the latter is isomorphic to the ring $\mathbb{Z}[\zeta]$ of integral elements of the field $\mathbb{Q}(\zeta)$ of p-th roots of unity (ζ a primitive p-th root of unity), and the map $\sigma \mapsto \zeta$ induces an isomorphism $\mathbb{Z}[G]/\mathbb{Z} \cdot N_G \cong \mathbb{Z}[\zeta]$. In $\mathbb{Z}[\zeta]$ we now have the well-known decomposition $p = (\zeta - 1)^{p-1} \cdot e$, e a unit, so that we can write
$$p = (\sigma - 1)^{p-1} \cdot \varepsilon, \quad \varepsilon \text{ unit in } \mathbb{Z}[G]/\mathbb{Z} \cdot N_G.$$

Since the endomorphism induced by ε is an automorphism on $I_G A$, we find $q_{0,\varepsilon}(I_G A) = 1$. If we now apply Lemma (6.8), we obtain
$$q_{0,p}(I_G A) = q_{0,D^{p-1}}(I_G A) \cdot q_{0,\varepsilon}(I_G A) = q_{0,D}(I_G A)^{p-1} = 1/q_{D,0}(I_G A)^{p-1}.$$

Since $N = N_G$ is the 0-endomorphism on $I_G A$, we also have
$$q_{0,p}(I_G A) = 1/q_{D,0}(I_G A)^{p-1} = 1/q_{D,N}(I_G A)^{p-1} = 1/h(I_G A)^{p-1}.$$

In combination with $(*)$, this implies

$$q_{0,p}(A^G) = h(A^G), q_{0,p}(I_G A) = 1/h(I_G A)^{p-1}, q_{0,p}(A) = q_{0,p}(A^G)/h(I_G A)^{p-1}.$$

On the other hand, the sequence $0 \to A^G \to A \to I_G A \to 0$ gives the formula

$$h(A)^{p-1} = h(A^G)^{p-1} \cdot h(I_G A)^{p-1},$$

and the claim $h(A)^{p-1} = q_{0,p}(A^G)^p/q_{0,p}(A)$ follows by substitution.

In global class field theory we will apply this theorem to certain unit groups, about which we only know that they are finitely generated of known rank. We show that this alone suffices to compute the Herbrand quotient; namely, from (6.9) we get the following theorem of C. CHEVALLEY:

(6.10) Theorem. *Let A be a finitely generated G-module, where G is a cyclic group of prime order p. If α (resp. β) denotes the rank of the abelian group A (resp. A^G), then the Herbrand quotient $h(A)$ is given by the formula*

$$h(A) = p^{(p \cdot \beta - \alpha)/(p-1)}.$$

Proof. We can decompose A into its torsion group A_0 and its torsion-free part A_1: $A = A_0 \oplus A_1$. It follows that $A^G = A_0^G \oplus A_1^G$. Since A is finitely generated, A_0 is a finite group, rank $A_1 =$ rank $A = \alpha$ and rank $A_1^G =$ rank $A^G = \beta$. Thus

$$h(A)^{p-1} = h(A_1)^{p-1} = q_{0,p}(A_1^G)^p/q_{0,p}(A_1),$$

where $q_{0,p}(A_1^G) = (A_1^G : pA_1^G) = p^\beta$ and $q_{0,p}(A_1) = (A_1 : pA_1) = p^\alpha$.

§ 7. Tate's Theorem

Many theorems in cohomology show that the vanishing of the cohomology groups in two consecutive dimensions implies the vanishing in all dimensions. One of the most important results of this type is the following **Theorem of Cohomological Triviality**.

(7.1) Theorem. *Let A be a G-module. If there is a dimension q_0 such that*

$$H^{q_0}(g, A) = H^{q_0+1}(g, A) = 0$$

for all subgroups $g \subseteq G$, then A has trivial cohomology [17].

We will reduce the general case to that of cyclic groups G, where the result is an immediate consequence of Theorem (6.1). It is clear that it suffices to prove the following claim:

---

[17] This means that $H^q(g, A) = 0$ for all $q \in \mathbb{Z}$ and all subgroups g of G.

If $H^{q_0}(g, A) = H^{q_0+1}(g, A) = 0$ for all subgroups $g \subseteq G$, then $H^{q_0-1}(g, A) = 0$ and $H^{q_0+2}(g, A) = 0$ for all subgroups $g \subseteq G$.

Moreover, by dimension shifting, it suffices to consider the case $q_0 = 1$. To see this, note that if the claim holds for $q_0 = 1$, then the isomorphism from (3.15)

$$H^{q-m}(g, A^m) \cong H^q(g, A)$$

implies $H^1(g, A^{q_0-1}) \cong H^{q_0}(g, A) = 0$ and $H^2(g, A^{q_0-1}) \cong H^{q_0+1}(g, A) = 0$. Hence $H^{q-(q_0-1)}(g, A^{q_0-1}) \cong H^q(g, A) = 0$ for all q.

Assume $H^1(g, A) = H^2(g, A) = 0$ for all subgroups $g \subseteq G$. We need to show

(∗) $\qquad H^0(g, A) = H^3(g, A) = 0 \quad$ for all subgroups $g \subseteq G$.

We prove this by induction on the order $|G|$ of G; the case $|G| = 1$ is trivial.

Thus we assume that we have proved (∗) for all proper subgroups g of G; it remains to show $H^0(G, A) = H^3(G, A) = 0$. This is clear if G is not a p-group; in this case all Sylow subgroups are proper subgroups, and (4.17) shows that $H^0(G, A) = H^3(G, A) = 0$.

We may therefore assume that G is a p-group. Then there exists a normal subgroup $H \subset G$ such that the quotient G/H is a cyclic group of prime order. By the induction assumption we have

$$H^0(H, A) = H^3(H, A) = 0 \quad \text{as well as} \quad H^1(H, A) = H^2(H, A) = 0,$$

and, using (4.6) and (4.7), we obtain the isomorphisms

$$\inf : H^q(G/H, A^H) \longrightarrow H^q(G, A) \quad \text{for } q = 1, 2, 3.$$

Now $H^1(G, A) = 0$ implies $H^1(G/H, A^H) = 0$, hence $H^3(G/H, A^H) = 0$ by (6.1), and so $H^3(G, A) = 0$.

Next $H^2(G, A) = 0$ implies $H^2(G/H, A^H) = 0$, hence $H^0(G/H, A^H) = 0$ (by (6.1)), which means $A^G = N_{G/H}A^H = N_{G/H}(N_H A) = N_G A$; here we have used $H^0(H, A) = 0$, i.e., $A^H = N_H A$. Thus $H^0(G, A) = 0$, which proves the theorem.

If A and B are G-modules, consider the cup product, i.e., the bilinear map

$$H^p(G, A) \times H^q(G, B) \xrightarrow{\cup} H^{p+q}(G, A \otimes B).$$

For a fixed element $a \in H^p(G, A)$, the map

$$a \cup : H^q(G, B) \longrightarrow H^{p+q}(G, A \otimes B), \ b \mapsto a \cup b \ \ (b \in H^q(G, B))$$

provides us with a whole family of maps. In the theorems below, we will use the cup product in this way.

From the Theorem of Cohomological Triviality we deduce the following result:

(7.2) Theorem. *Let A be a G-module with the following properties: For each subgroup $g \subseteq G$ we have*

   I. $H^{-1}(g, A) = 0$,
   II. $H^0(g, A)$ *is a cyclic group of order* $|g|$.

If a generates the group $H^0(G, A)$, then the cup product map

$$a \cup \ : H^q(G, \mathbb{Z}) \longrightarrow H^q(G, A)$$

is an isomorphism for all $q \in \mathbb{Z}$.

Proof. The module A itself is not suitable for the proof, since we need to use the injectivity of the map $\mathbb{Z} \to A$, $n \mapsto na_0$ ($a_0 + N_G A = a$), which induces the cup product above for the case $q = 0$ (cf. (5.2)). Hence we replace A with

$$B = A \oplus \mathbb{Z}[G]$$

which we can do without changing the cohomology groups. In fact, if $i : A \to B$ is the canonical injection onto the first component of B, then the induced map

$$\bar{i} : H^q(g, A) \longrightarrow H^q(g, B)$$

is an isomorphism, because $\mathbb{Z}[G]$ is cohomologically trivial. Now we choose an $a_0 \in A^G$ such that $a = a_0 + N_G A$ is a generator of $H^0(G, A)$. Then the map

$$f : \mathbb{Z} \longrightarrow B \quad \text{with} \quad n \longmapsto a_0 \cdot n + N_G \cdot n.$$

is injective, because of the second term $N_G \cdot n$, and induces the homomorphism

$$\bar{f} : H^q(g, \mathbb{Z}) \longrightarrow H^q(g, B).$$

Using (5.2), we see that the diagram

$$
\begin{array}{ccc}
H^q(G, \mathbb{Z}) & \xrightarrow{\ a\cup\ } & H^q(G, A) \\
& \searrow{\scriptstyle f} & \downarrow{\scriptstyle \bar{i}} \\
& & H^q(G, B)
\end{array}
$$

commutes, thus it suffices to show $\bar{f}$ is bijective. This follows easily from (7.1): Since the map $f : \mathbb{Z} \to B$ is injective, there is an exact sequence of G-modules

$$(*) \qquad\qquad 0 \longrightarrow \mathbb{Z} \xrightarrow{\ f\ } B \longrightarrow C \longrightarrow 0.$$

Now $H^{-1}(g, B) = H^{-1}(g, A) = 0$ and $H^1(g, \mathbb{Z}) = 0$ for all $g \subseteq G$, which implies that the corresponding exact cohomology sequence has the form

$$0 \longrightarrow H^{-1}(g, C) \longrightarrow H^0(g, \mathbb{Z}) \xrightarrow{\ \bar{f}\ } H^0(g, B) \longrightarrow H^0(g, C) \longrightarrow 0.$$

If $q = 0$, then $\bar{f}$ is clearly an isomorphism, thus $H^{-1}(g, C) = H^0(g, C) = 0$. By (7.1) the vanishing of two consecutive cohomology groups implies $H^q(g, C) = 0$ for all q, and it follows from the exact cohomology sequence associated with $(*)$ that $\bar{f} : H^q(G, \mathbb{Z}) \to H^q(G, B)$ is bijective for all q, as claimed.

From (7.2) we obtain the following, very important result:

(7.3) Tate's Theorem. *Assume that A is a G-module with the following properties: For each subgroup $g \subseteq G$ we have I. $H^1(g, A) = 0$ and II. $H^2(g, A)$ is cyclic of order $|g|$. If a generates the group $H^2(G, A)$, then the map*

$$a \cup \; : H^q(G, \mathbb{Z}) \longrightarrow H^{q+2}(G, A)$$

is an isomorphism.

Addendum: *If a generates the group $H^2(G, A)$, then $\operatorname{res} a \in H^2(g, A)$ generates the group $H^2(g, A)$. Thus we also have the isomorphism*

$$\operatorname{res} a \cup \; : H^q(g, \mathbb{Z}) \longrightarrow H^{q+2}(g, A).$$

Proof.   Consider the isomorphism $\delta^2 : H^q(g, A^2) \to H^{q+2}(g, A)$ from (3.15). The assumptions I. and II. imply that $H^{-1}(g, A^2) = 0$, and that $H^0(g, A^2)$ is cyclic of order $|g|$. Furthermore, the generator $a \in H^2(G, A)$ is the image of the generator $\delta^{-2}a \in H^0(G, A^2)$. It follows from (5.1) that the diagram

$$
\begin{array}{ccc}
H^q(G, \mathbb{Z}) & \xrightarrow{\;\delta^{-2}a\,\cup\;} & H^q(G, A^2) \\[4pt]
{\scriptstyle \mathrm{id}}\downarrow & & \downarrow{\scriptstyle \delta^2} \\[4pt]
H^q(G, \mathbb{Z}) & \xrightarrow{\;a\,\cup\;} & H^{q+2}(G, A) \,,
\end{array}
$$

commutes. Since $\delta^{-2}a \cup$ is bijective by (7.2), the map $a \cup$ is bijective as well.

As for the addendum: Since $\operatorname{cor} \circ \operatorname{res} a = (G : g) \cdot a$, the order of the element $\operatorname{res} a \in H^2(g, A)$ is divisible by $|g|$, hence $\operatorname{res} a$ generates $H^2(g, A)$ by II.

Tate's Theorem can be generalized considerably. For example, we may replace the condition "for all subgroups $g \subseteq G$" by "for all p-Sylow subgroups". In addition, the shifting from q to $q + 2$ by two dimensions may be extended (under suitable assumptions) to general dimensions. Finally, the G-module $\mathbb{Z}$ may be replaced by more general modules[18]. We do not discuss this in detail, since the form of Tate's Theorem presented here suffices for most applications.

For class field theory, the case $q = -2$ is particularly important. In this case, Tate's Theorem yields a canonical isomorphism between the abelianization $G^{\mathrm{ab}} \cong H^{-2}(G, \mathbb{Z})$ of G and the norm residue group $A^G / N_G A = H^0(G, A)$:

$$G^{\mathrm{ab}} \longrightarrow A^G / N_G A.$$

This canonical isomorphism is the abstract formulation of the main theorem of class field theory, the so-called "reciprocity law". For this reason, one can consider Tate's Theorem as the foundation for a purely group theoretically formulated abstract version of class field theory. In the next part, we will develop this idea in detail.

---

[18] Cf. [42, IX, §8, Th. 13, p. 156].

Part II

Local Class Field Theory

J. Neukirch, *Class Field Theory*, DOI 10.1007/978-3-642-35437-3_2,
© Springer-Verlag Berlin Heidelberg 2013

§ 1. Abstract Class Field Theory

Local and global class field theory, as well as a series of further theories for which the name class field theory is similarly justified, have the following principle in common. All of these theories involve a canonical bijective correspondence between the abelian extensions of a field K and certain subgroups of a corresponding module A_K associated with the field K. This correspondence has the property that if the subgroup $I \subseteq A_K$ corresponds to an abelian field extension $L|K$ (the "class field associated with I"), then there exists a canonical isomorphism between the Galois group $G_{L|K}$ and the factor group A_K/I. This so-called reciprocity law is the main theorem of class field theory.

This main theorem can be traced back to a common system of axioms for the concrete theories mentioned above which essentially consists of the assumptions in Tate's Theorem (cf. I, §7); in fact one can view Tate's Theorem itself as the abstract version of the main theorem of class field theory. The notion of a **class formation** is based on this idea. It separates the purely group theoretic machinery, which is characteristic of class field theory, from the specific considerations of field theory, and gives in an easily comprehensible and elegant way information about the goal and function of the theory.

Let G be a profinite group, i.e., a compact group with the normal-subgroup topology[1]. We may think of G as the Galois group (endowed with the Krull topology) of an infinite Galois field extension, although the abstract notions in this section do not use this interpretation. The open subgroups of G are precisely the closed subgroups of finite index. In fact, the complement of an open subgroup is the union of (open) cosets, thus open, and since G is compact, finitely many of these cosets cover the group G, hence the index is finite. Conversely, a closed subgroup of finite index is open, because it is the union of finitely many cosets, hence its complement is closed.

Given a profinite group G, we consider the family $\{G_K \mid K \in X\}$ of all open subgroups of G, i.e., the closed subgroups of finite index. We label each such subgroup with the index K, and call these indices "fields".

The "field" K_0 with $G_{K_0} = G$ is called the base field. If $G_L \subseteq G_K$, we write formally $K \subseteq L$, and define the degree of such an extension $L|K$ as

$$[L : K] = (G_K : G_L).$$

The extension $L|K$ is called normal if $G_L \subseteq G_K$ is a normal subgroup of G_K. If $L|K$ is normal, then the Galois group of $L|K$ is defined as the quotient group

$$G_{L|K} = G_K/G_L.$$

---

[1] We refer to [9], [28], [41] for the theory of profinite groups.

An extension $L|K$ is called cyclic, abelian, solvable, etc., if its Galois group $G_{L|K} = G_K/G_L$ is cyclic, abelian, solvable, etc. We define the intersection and the compositum of such fields K_i by setting

$$K = \bigcap_{i=1}^{n} K_i, \text{ if } G_K \text{ is (topologically) generated by the } G_{K_i} \text{ in } G; \text{ and}$$

$$K = \prod_{i=1}^{n} K_i, \text{ if } G_K = \bigcap_{i=1}^{n} G_{K_i}.$$

If $G_{L'} = \sigma G_L \sigma^{-1}$ for $\sigma \in G$, then we write $L' = \sigma L$, and we call two extensions $L|K$ and $L'|K$ conjugate in case $L' = \sigma L$ for some $\sigma \in G_K$. With these notions, we obtain for each profinite group G a formal Galois theory.

In the following we consider modules A on which a profinite group G acts. In this context it is important to keep the topological structure on G in mind. The action of G on A should be in a certain sense continuous. More precisely, it should satisfy one of the following, equivalent conditions:

(i) The map $G \times A \to A$ with $(\sigma, a) \mapsto \sigma a$ is continuous[2],
(ii) For each $a \in A$ the stabilizer $\{\sigma \in G \mid \sigma a = a\}$ is open in G,
(iii) $A = \bigcup_U A^U$, where U runs through all the open subgroups of G.

(1.1) Definition. *If G is a profinite group and A is a G-module satisfying the equivalent conditions (i)-(iii) above, the pair (G, A) is called a* **formation**.

If G is the Galois group of a (infinite) Galois extension $N|K$, then G acts on the multiplicative group $N^\times$ of the field N, and the pair $(G, N^\times)$ is a formation. It is precisely this example that comes into play in local class field theory, and one may use it as an orientation for what follows.

Let (G, A) be a formation. In the following we think of the module A as multiplicatively written. Let $\{G_K \mid K \in X\}$ be the family of open subgroups of G, indexed by the set of fields X. For each field $K \in X$ we consider the fixed module associated with K, i.e.,

$$A_K = A^{G_K} = \{a \in A \mid \sigma a = a \text{ for all } \sigma \in G_K.\}$$

In the class field theory example mentioned above, we obviously have $A_K = K^\times$. If $K \subseteq L$, then $A_K \subseteq A_L$.

If $L|K$ is a normal extension, then A_L is a $G_{L|K}$-module. When we call the pair (G, A) a formation, we basically mean by this the formation of these normal extensions $L|K$, together with the $G_{L|K}$-modules A_L.

---

[2] Here A is interpreted as a discrete module.

We consider now for each normal extension $L|K$ the cohomology groups of the $G_{L|K}$-module A_L. For simplicity of notation, we set

$$H^q(L|K) = H^q(G_{L|K}, A_L).$$

If $N \supseteq L \supseteq K$ is a tower of normal extensions of K, we have inclusions $G_N \subseteq G_L \subseteq G_K$ with G_N and G_L normal in G_K, and cohomology theory yields the homomorphism

$$H^q(G_{L|K}, A_L) = H^q(G_{L|K}, A_N^{G_{N|L}}) \xrightarrow{\text{inf}} H^q(G_{N|K}, A_N),$$

in other words

$$H^q(L|K) \xrightarrow{\text{inf}_N} H^q(N|K) \qquad \text{for } q \geq 1.$$

In addition, we also have the restriction and corestriction maps

$$H^q(G_{N|K}, A_N) \xrightarrow{\text{res}} H^q(G_{N|L}, A_N) \text{ and } H^q(G_{N|L}, A_N) \xrightarrow{\text{cor}} H^q(G_{N|K}, A_N),$$

that is, for every integer q homomorphims

$$H^q(N|K) \xrightarrow{\text{res}_L} H^q(N|L) \text{ and } H^q(N|L) \xrightarrow{\text{cor}_K} H^q(N|K).$$

Here we need only need to assume that $N|K$ is normal. If both N and L are normal, then the sequence

$$1 \longrightarrow H^q(L|K) \xrightarrow{\text{inf}_N} H^q(N|K) \xrightarrow{\text{res}_L} H^q(N|L)$$

is exact for $q = 1$, and exact for $q > 1$ if $H^i(N|L) = 1$ for $i = 1, \ldots, q-1$ (cf. I, (4.7)).

If $L|K$ is normal and $\sigma \in G$, then

$$\tau G_L \longmapsto \sigma\tau\sigma^{-1} G_{\sigma L}$$

defines an isomorphism between $G_{L|K}$ and $G_{\sigma L|\sigma K}$, and

$$a \longmapsto \sigma a$$

an isomorphism between A_L and $A_{\sigma L}$. Since $(\sigma\tau\sigma^{-1}G_{\sigma L})\sigma a = \sigma(\tau G_L)a$, these isomorphisms are compatible, and we obtain an equivalence between the $G_{L|K}$-module A_L and the $G_{\sigma L|\sigma K}$-module $A_{\sigma L}$. Thus every $\sigma \in G$ yields an isomorphism

$$H^q(L|K) \xrightarrow{\sigma^*} H^q(\sigma L|\sigma K).$$

Using the equivalence of the modules A_L and $A_{\sigma L}$, it is easy to see that the isomorphism σ^* commutes with inflation, restriction and corestriction.

We call a formation (G, A) a **field formation** when for each normal extension the first cohomology group vanishes:

$$H^1(L|K) = 1.$$

In a field formation the sequence

$$1 \longrightarrow H^2(L|K) \xrightarrow{\text{inf}_N} H^2(N|K) \xrightarrow{\text{res}_L} H^2(N|L) \qquad (N \supseteq L \supseteq K)$$

is always exact (cf. I, (4.7)). We will soon see that the example mentioned above, where G is the Galois group of a Galois field extension and A the multiplicative group of the extension field, represents such a field formation. If $N \supseteq L \supseteq K$ are normal extensions, then we can always think of the group $H^2(L|K)$ as embedded in the group $H^2(N|K)$, since the inflation map

$$H^2(L|K) \xrightarrow{\ \inf_N\ } H^2(N|K)$$

is injective. The presentation of class field theory will become formally especially simple, if we take this identification one step further. If L ranges over the normal extensions of K, then the groups $H^2(L|K)$ form a direct system of groups with respect to the inflation maps. Taking the direct limit

$$H^2(\ |K) = \varinjlim_L H^2(L|K)$$

we obtain a group $H^2(\ |K)$ in which all the groups $H^2(L|K)$ are embedded via the injective inflation maps. If we identify these groups with their images under this embedding, then $H^2(L|K)$ become subgroups of $H^2(\ |K)$, and

$$H^2(\ |K) = \bigcup_L H^2(L|K).$$

In particular, if $N \supseteq L \supseteq K$ is a tower of normal extensions of K, we have so

$$H^2(L|K) \subseteq H^2(N|K) \subseteq H^2(\ |K).$$

We strongly emphasize that the inflation maps are to be interpreted as inclusions here. An element of $H^2(N|K)$ is regarded as an element of $H^2(L|K)$ if it is the inflation of an element of $H^2(L|K)$.

Remark. Let G_K be a profinite group and let A be a G_K-module. Exactly as for finite groups, we can define cohomology groups $H^q(G_K, A)$, $q \geq 0$ by taking as cochains the **continuous** maps $G_K \times \ldots \times G_K \to A$. Then (cf. [41])

$$H^q(G_K, A) \cong H^q(\ |K) = \varinjlim_L H^q(L|K).$$

Given any extension $K'|K$ of K, we obtain a canonical homomorphism

$$H^2(\ |K) \xrightarrow{\ \operatorname{res}_{K'}\ } H^2(\ |K').$$

In fact, if $c \in H^2(\ |K)$, then there is an extension $L \supseteq K' \supseteq K$, so that c is contained in the group $H^2(L|K)$, hence the restriction map

$$H^2(L|K) \xrightarrow{\ \operatorname{res}_{K'}\ } H^2(L|K')$$

defines an element

$$\operatorname{res}_{K'} c \in H^2(L|K') \subseteq H^2(\ |K').$$

The map $c \mapsto \mathrm{res}_{K'}c$ is independent of the choice of the field $L \supseteq K'$; this follows from the trivial fact that restriction commutes with inflation, which is interpreted as inclusion. The restriction of the map $H^2(\ |K) \xrightarrow{\mathrm{res}_{K'}} H^2(\ |K')$ to the group $H^2(L|K)$ $(L \supseteq K' \supseteq K)$ gives back the usual restriction map

$$H^2(L|K) \xrightarrow{\mathrm{res}_{K'}} H^2(L|K').$$

From this we immediately obtain

(1.2) Proposition. *Let (G, A) be a field formation. If $K'|K$ is normal, then*

$$1 \longrightarrow H^2(K'|K) \xrightarrow{\mathrm{incl}} H^2(\ |K) \xrightarrow{\mathrm{res}_{K'}} H^2(\ |K')$$

is exact.

The fundamental assertion in both local and global class field theory is the existence of a canonical isomorphism, the so-called "reciprocity map"

$$G_{L|K}^{\mathrm{ab}} \cong A_K/N_{L|K}A_L$$

for every normal extension $L|K$, where $G_{L|K}^{\mathrm{ab}}$ is the abelianization of $G_{L|K}$ and $N_{L|K}A_L = N_{G_{L|K}}A_L$ is the norm group of A_L. Because of Tate's Theorem, we can force the existence of such an isomorphism *in abstracto* by imposing the following conditions on our formation (G, A): If $L|K$ is any extension, then

$$\text{I. } H^1(L|K) = 1 \quad \text{and} \quad \text{II. } H^2(L|K) \text{ is cyclic of order } [L:K].$$

If this holds, then the cup product with a generator of $H^2(L|K)$ gives an isomorphism

$$G_{L|K}^{\mathrm{ab}} \cong A_K/N_{L|K}A_L.$$

However, there is a certain arbitrariness to this isomorphism, since it depends on the choice of the generator of $H^2(L|K)$. In order to get a "canonical" reciprocity map, we replace II. by the condition that there is an isomorphism between $H^2(L|K)$ and the cyclic group $\frac{1}{[L:K]}\mathbb{Z}/\mathbb{Z}$, the so-called "invariant map", which uniquely determines the element $u_{L|K} \in H^2(L|K)$ with image $\frac{1}{[L:K]} + \mathbb{Z}$. The crucial point here is that this element $u_{L|K}$ remains "correct" when passing to extension fields and subfields, which we ensure by imposing certain compatibility conditions on the invariant map.

These considerations lead us to the following

(1.3) Definition. *A formation (G, A) is be called a **class formation** if it satisfies the following axioms:*

Axiom I. $H^1(L|K) = 1$ *for every normal extension $L|K$ (field formation).*

Axiom II. *For every normal extension $L|K$ there exists an isomorphism*

$$\mathrm{inv}_{L|K} : H^2(L|K) \longrightarrow \tfrac{1}{[L:K]}\mathbb{Z}/\mathbb{Z},$$

*the **invariant map**, with the following properties:*

a) If $N \supseteq L \supseteq K$ is a tower of normal extensions, then

$$\operatorname{inv}_{L|K} = \operatorname{inv}_{N|K}|_{H^2(L|K)}.$$

b) If $N \supseteq L \supseteq K$ is a tower of extensions with $N|K$ normal, then

$$\operatorname{inv}_{N|L} \circ \operatorname{res}_L = [L : K] \cdot \operatorname{inv}_{N|K}.$$

Remark. Formula II b) becomes almost obvious if one replaces it by the commutative diagram

$$
\begin{array}{ccc}
H^2(N|K) & \xrightarrow{\operatorname{inv}_{N|K}} & \frac{1}{[N:K]}\mathbb{Z}/\mathbb{Z} \\
{\scriptstyle \operatorname{res}_L}\downarrow & & \downarrow{\scriptstyle \cdot[L:K]} \\
H^2(N|L) & \xrightarrow{\operatorname{inv}_{N|L}} & \frac{1}{[N:L]}\mathbb{Z}/\mathbb{Z}.
\end{array}
$$

The extension property II a) of the invariant map implies that if $H^2(\ |K) = \bigcup_L H^2(L|K)$, then there is an injective homomorphism

$$\operatorname{inv}_K : H^2(\ |K) \longrightarrow \mathbb{Q}/\mathbb{Z}.$$

For this map we obtain from formula II b) the following relation: If $L|K$ is an arbitrary extension of K, then

$$\operatorname{inv}_L \circ \operatorname{res}_L = [L : K] \cdot \operatorname{inv}_K,$$

where res_L is the homomorphism $H^2(\ |K) \xrightarrow{\operatorname{res}_L} H^2(\ |L)$ defined on p. 66. Conversely, we recover from this equality formula II b), since $\operatorname{inv}_{N|L}$ (resp. $\operatorname{inv}_{N|K}$) is the restriction of inv_L (resp. inv_K) to $H^2(N|L)$ (resp. $H^2(N|K)$).

Together with the formulas of Axiom II, we immediately obtain additional formulas for the corestriction map cor and the conjugate map σ^* (cf. p. 65).

(1.4) Proposition. Let $N \supseteq L \supseteq K$ be extensions with $N|K$ normal. Then

a) $\operatorname{inv}_{N|K} c = \operatorname{inv}_{L|K} c,$ if $L|K$ is normal
  and $c \in H^2(L|K) \subseteq H^2(N|K),$

b) $\operatorname{inv}_{N|L}(\operatorname{res}_L c) = [L : K] \cdot \operatorname{inv}_{N|K} c,$ for $c \in H^2(N|K),$

c) $\operatorname{inv}_{N|K}(\operatorname{cor}_K c) = \operatorname{inv}_{N|L} c,$ for $c \in H^2(N|L),$

d) $\operatorname{inv}_{\sigma N|\sigma K}(\sigma^* c) = \operatorname{inv}_{N|K} c,$ for $c \in H^2(N|K)$ and $\sigma \in G.$

Proof. a) and b) are just restatements of the formulas in Axiom II.

c) The commutative diagram on p. 68 immediately implies that the map $H^2(N|K) \xrightarrow{\operatorname{res}_L} H^2(N|L)$ is surjective. Hence for every $c \in H^2(N|L)$ we have $c = \operatorname{res}_L \tilde{c},\ \tilde{c} \in H^2(N|K),$ and so $\operatorname{cor}_K c = \operatorname{cor}_K(\operatorname{res}_L \tilde{c}) = \tilde{c}^{[L:K]}$ (cf. I, (4.14)). Thus, by b), $\operatorname{inv}_{N|K}(\operatorname{cor}_K c) = [L : K] \cdot \operatorname{inv}_{N|K}(\tilde{c}) = \operatorname{inv}_{N|L}(\operatorname{res}_L \tilde{c}) = \operatorname{inv}_{N|L} c.$

d) Let $\widetilde{N}$ be a normal extension of the base field K_0 corresponding to the group G such that $\widetilde{N}$ contains N. Then $\sigma\widetilde{N} = \widetilde{N}$, i.e., the map $a \mapsto \sigma a$ defines a $G_{\widetilde{N}|K_0}$-automorphism of the $G_{\widetilde{N}|K_0}$-module $A_{\widetilde{N}}$ such that

$$\sigma^* : H^2(\widetilde{N}|K_0) \longrightarrow H^2(\widetilde{N}|K_0)$$

is the identity map on $H^2(\widetilde{N}|K_0)$. Since σ^* commutes with inflation (inclusion) and corestriction (cf. p. 65), we have by a) and c) for every $c \in H^2(N|K)$

$$\mathrm{inv}_{\sigma N|\sigma K}(\sigma^* c) = \mathrm{inv}_{\widetilde{N}|\sigma K}(\sigma^* c) = \mathrm{inv}_{\widetilde{N}|K_0}(\mathrm{cor}_{K_0}(\sigma^* c)) =$$
$$\mathrm{inv}_{\widetilde{N}|K_0}(\sigma^* \mathrm{cor}_{K_0} c) = \mathrm{inv}_{\widetilde{N}|K_0}(\mathrm{cor}_{K_0} c) = \mathrm{inv}_{\widetilde{N}|K} c = \mathrm{inv}_{N|K} c.$$

Now we can distinguish a "canonical" generator in each group $H^2(L|K)$.

(1.5) Definition. *Let $L|K$ be a normal extension. The uniquely determined element $u_{L|K} \in H^2(L|K)$ such that*

$$\mathrm{inv}_{L|K}(u_{L|K}) = \tfrac{1}{[L:K]} + \mathbb{Z}$$

*is called the **fundamental class** of $L|K$.*

From the behavior of the invariant map described in Proposition (1.4), we see how the fundamental classes of different field extensions are related.

(1.6) Proposition. *Let $N \supseteq L \supseteq K$ be extensions with $N|K$ normal. Then*

a) $u_{L|K} = (u_{N|K})^{[N:L]}$, *if $L|K$ is normal,*
b) $\mathrm{res}_L(u_{N|K}) = u_{N|L}$,
c) $\mathrm{cor}_K(u_{N|L}) = (u_{N|K})^{[L:K]}$,
d) $\sigma^*(u_{N|K}) = u_{\sigma N|\sigma K}$ *for $\sigma \in G$.*

Proof. Since two cohomology classes are equal if they have the same invariants, the proposition follows from

a) $\mathrm{inv}_{N|K}((u_{N|K})^{[N:L]}) = [N:L] \cdot \mathrm{inv}_{N|K}(u_{N|K}) = \frac{[N:L]}{[N:K]} + \mathbb{Z} = \frac{1}{[L:K]} + \mathbb{Z}$
$= \mathrm{inv}_{L|K}(u_{L|K}) = \mathrm{inv}_{N|K}(u_{L|K}),$

b) $\mathrm{inv}_{N|L}(\mathrm{res}_L(u_{N|K})) = [L:K] \cdot \mathrm{inv}_{N|K}(u_{N|K}) = \frac{[L:K]}{[N:K]} + \mathbb{Z} = \frac{1}{[N:L]} + \mathbb{Z}$
$= \mathrm{inv}_{N|L}(u_{N|L}),$

c) $\mathrm{inv}_{N|K}(\mathrm{cor}_K(u_{N|L})) = \mathrm{inv}_{N|L}(u_{N|L}) = \frac{1}{[N:L]} + \mathbb{Z} = \frac{[L:K]}{[N:K]} + \mathbb{Z}$
$= [L:K] \cdot \mathrm{inv}_{N|K}(u_{N|K}) = \mathrm{inv}_{N|K}((u_{N|K})^{[L:K]}),$

d) $\mathrm{inv}_{\sigma N|\sigma K}(\sigma^* u_{N|K}) = \mathrm{inv}_{N|K}(u_{N|K}) = \frac{1}{[N:K]} + \mathbb{Z} = \frac{1}{[\sigma N:\sigma K]} + \mathbb{Z}$
$= \mathrm{inv}_{\sigma N|\sigma K}(u_{\sigma N|\sigma K}).$

Now we apply Tate's Theorem, I, (7.3) to obtain the main theorem of class formations.

(1.7) Main Theorem. *Let $L|K$ be a normal extension. Then the map*

$$u_{L|K} \cup \, : H^q(G_{L|K}, \mathbb{Z}) \longrightarrow H^{q+2}(L|K)$$

given by the cup product with the fundamental class $u_{L|K} \in H^2(L|K)$ is an isomorphism in all dimensions q.

For $q = 1, 2$ we immediately obtain

(1.8) Corollary. $H^3(L|K) = 1$ *and* $H^4(L|K) \cong \chi(G_{L|K})$.

Proof. We have $H^3(L|K) \cong H^1(G_{L|K}, \mathbb{Z}) = \operatorname{Hom}(G_{L|K}, \mathbb{Z}) = 0$, and $H^4(L|K) \cong H^2(G_{L|K}, \mathbb{Z}) \cong H^1(G_{L|K}, \mathbb{Q}/\mathbb{Z}) = \operatorname{Hom}(G_{L|K}, \mathbb{Q}/\mathbb{Z}) = \chi(G_{L|K})$; here the second isomorphism $H^2(G_{L|K}, \mathbb{Z}) \cong H^1(G_{L|K}, \mathbb{Q}/\mathbb{Z})$ follows from the exact cohomology sequence associated with $0 \to \mathbb{Z} \to \mathbb{Q} \to \mathbb{Q}/\mathbb{Z} \to 0$, using that the $G_{L|K}$-module $\mathbb{Q}$ is cohomologically trivial (since $\mathbb{Q}$ is a uniquely divisible group).

Since we do not have a concrete interpretation of the groups $H^q(L|K)$ in case $q = 3, 4$, or generally for all cohomology groups of higher dimension [3], Corollary (1.8) has no immediate concrete application. However, if $q = -2$, then we have such an interpretation because of the canonical isomorphisms

$$G_{L|K}^{\mathrm{ab}} \cong H^{-2}(G_{L|K}, \mathbb{Z}) \quad \text{and} \quad H^0(L|K) = A_K/N_{L|K}A_L.$$

Thus we obtain the following important **general reciprocity law:**

(1.9) Theorem. *Let $L|K$ be a normal extension. Then the cup product map*

$$u_{L|K} \cup : H^{-2}(G_{L|K}, \mathbb{Z}) \longrightarrow H^0(L|K)$$

yields a canonical isomorphism

$$\theta_{L|K} : G_{L|K}^{\mathrm{ab}} \longrightarrow A_K/N_{L|K}A_L$$

between the abelianization of the Galois group and the norm residue group of the module.

The isomorphism $\theta_{L|K}$ in Theorem (1.9) is called the **Nakayama map.** Using I, (5.8) we can give an explicit description of this map as follows:

---

[3] Recently, however, the case $q = -3$ has been found to have a beautiful application in connection with the solution of the "class field tower problem" (cf. [14] and [41], Ch. I, 4.4).

If u is a 2-cocycle representing the fundamental class $u_{L|K}$, then we have

$$\theta_{L|K}(\sigma G'_{L|K}) = \left[\prod_{\tau \in G_{L|K}} u(\tau, \sigma) \right] \cdot N_{L|K} A_L$$

for all $\sigma G'_{L|K} \in G^{\mathrm{ab}}_{L|K} = G_{L|K}/G'_{L|K}$.

Despite this description, it turns out that the inverse isomorphism of $\theta_{L|K}$,

$$A_K/N_{L|K} A_L \longrightarrow G^{\mathrm{ab}}_{L|K},$$

which is also called the **reciprocity isomorphism**, is often more accessible, and also more important, in particular for local and global class field theory. It induces a homomorphism from A_K onto $G^{\mathrm{ab}}_{L|K}$ with kernel $N_{L|K} A_L$. This homomorphism is called the **norm residue symbol** $(\ , L|K)$. Hence we have the exact sequence

$$1 \longrightarrow N_{L|K} A_L \longrightarrow A_K \xrightarrow{(\ , L|K)} G^{\mathrm{ab}}_{L|K} \longrightarrow 1 \,,$$

and an element $a \in A_K$ is a norm if and only if $(a, L|K) = 1$.

The following lemma establishes a relation between the norm residue symbol $(\ , L|K)$ and the invariant map $\mathrm{inv}_{L|K}$, which will be useful later.

(1.10) Lemma. *Let $L|K$ be a normal extension, $a \in A_K$, and $\bar{a} = a \cdot N_{L|K} A_L \in H^0(L|K)$. If $\chi \in \chi(G^{\mathrm{ab}}_{L|K}) = H^1(G_{L|K}, \mathbb{Q}/\mathbb{Z})$ is a character, then*

$$\chi((a, L|K)) = \mathrm{inv}_{L|K}(\bar{a} \cup \delta\chi) \in \tfrac{1}{[L:K]}\mathbb{Z}/\mathbb{Z}.$$

In this formula, the symbol $\delta\chi$ denotes the image of χ under the isomorphism

$$H^1(G_{L|K}, \mathbb{Q}/\mathbb{Z}) \xrightarrow{\delta} H^2(G_{L|K}, \mathbb{Z}),$$

which is obtained from the exact sequence

$$0 \longrightarrow \mathbb{Z} \longrightarrow \mathbb{Q} \longrightarrow \mathbb{Q}/\mathbb{Z} \longrightarrow 0.$$

Note that the above formula provides us with a characterization of the norm residue symbol $(a, L|K)$ in terms of the invariant map, since an element of $G^{\mathrm{ab}}_{L|K}$ is uniquely determined by its values under all characters.

Proof. To simplify notation we set

$$\sigma_a = (a, L|K) \in G^{\mathrm{ab}}_{L|K} \cong H^{-2}(G_{L|K}, \mathbb{Z})$$

and denote by $\bar{\sigma}_a$ the element in $H^{-2}(G_{L|K}, \mathbb{Z})$ corresponding to σ_a under the above isomorphism. By definition of the norm residue symbol we have

$$\bar{a} = u_{L|K} \cup \bar{\sigma}_a \in H^0(G_{L|K}, A_L).$$

Since the cup product is associative and commutes with the δ-map, we obtain

$$\bar{a} \cup \delta\chi = (u_{L|K} \cup \bar{\sigma}_a) \cup \delta\chi = u_{L|K} \cup (\bar{\sigma}_a \cup \delta\chi) = u_{L|K} \cup \delta(\bar{\sigma}_a \cup \chi).$$

By I, (5.7) we further have

$$\bar{\sigma}_a \cup \chi = \chi(\sigma_a) = \tfrac{r}{n} + \mathbb{Z} \in \tfrac{1}{n}\mathbb{Z}/\mathbb{Z} = H^{-1}(G_{L|K}, \mathbb{Q}/\mathbb{Z}),$$

where $n = [L : K]$. Hence taking $\delta : H^{-1}(G_{L|K}, \mathbb{Q}/\mathbb{Z}) \to H^0(G_{L|K}, \mathbb{Z})$ gives

$$\delta(\chi(\sigma_a)) = n(\tfrac{r}{n} + \mathbb{Z}) = r + n\mathbb{Z} \in H^0(G_{L|K}, \mathbb{Z}) = \mathbb{Z}/n\mathbb{Z},$$

and therefore

$$\bar{a} \cup \delta\chi = u_{L|K} \cup (r + n\mathbb{Z}) = u_{L|K}^r.$$

From this we get

$$\mathrm{inv}_{L|K}(\bar{a} \cup \delta\chi) = r \cdot \mathrm{inv}_{L|K}(u_{L|K}) = \tfrac{r}{n} + \mathbb{Z} = \chi(\sigma_a).$$

The conditions on the behavior of the invariant map under the inflation (inclusion) and restriction maps in Axiom II of the definition of a class formation already determines how the norm residue symbol behaves when passing to extension and subfields. We summarize this in the following theorem.

(1.11) Theorem. *Let $N \supseteq L \supseteq K$ be a tower of extensions of K with $N|K$ normal. Then the following diagrams are commutative:*

a)

$$\begin{array}{ccc} A_K & \xrightarrow{\ (\ ,N|K)\ } & G_{N|K}^{\mathrm{ab}} \\ {\scriptstyle \mathrm{id}}\downarrow & & \downarrow{\scriptstyle \pi} \\ A_K & \xrightarrow{\ (\ ,L|K)\ } & G_{L|K}^{\mathrm{ab}} \end{array}$$

hence $(a, L|K) = \pi(a, N|K) \in G_{L|K}^{\mathrm{ab}}$ for $a \in A_K$, if $L|K$ is also normal (in addition to $N|K$). Here π is the canonical projection of $G_{N|K}^{\mathrm{ab}}$ onto $G_{L|K}^{\mathrm{ab}}$.

b)

$$\begin{array}{ccc} A_K & \xrightarrow{\ (\ ,N|K)\ } & G_{N|K}^{\mathrm{ab}} \\ {\scriptstyle \mathrm{incl}}\downarrow & & \downarrow{\scriptstyle \mathrm{Ver}} \\ A_L & \xrightarrow{\ (\ ,N|L)\ } & G_{N|L}^{\mathrm{ab}} \end{array}$$

hence $(a, N|L) = \mathrm{Ver}(a, N|K) \in G_{N|L}^{\mathrm{ab}}$ for $a \in A_K$. Recall that the Verlagerung (transfer) is induced from $H^{-2}(G_{N|K}, \mathbb{Z}) \xrightarrow{\mathrm{res}} H^{-2}(G_{N|L}, \mathbb{Z})$.

c)

$$
\begin{array}{ccc}
A_L & \xrightarrow{(\ ,N|L)} & G^{\mathrm{ab}}_{N|L} \\
{\scriptstyle N_{L|K}} \Big\downarrow & & \Big\downarrow {\scriptstyle \kappa} \\
A_K & \xrightarrow{(\ ,N|K)} & G^{\mathrm{ab}}_{N|K}
\end{array}
$$

hence $(N_{L|K}a, N|K) = \kappa(a, N|L) \in G^{\mathrm{ab}}_{N|K}$ for $a \in A_L$, where κ is the canonical homomorphism from $G^{\mathrm{ab}}_{N|L}$ into $G^{\mathrm{ab}}_{N|K}$.

d)

$$
\begin{array}{ccc}
A_K & \xrightarrow{(\ ,N|K)} & G^{\mathrm{ab}}_{N|K} \\
{\scriptstyle \sigma} \Big\downarrow & & \Big\downarrow {\scriptstyle \sigma^*} \\
A_{\sigma K} & \xrightarrow{(\ ,\sigma N|\sigma K)} & G^{\mathrm{ab}}_{\sigma N|\sigma K}
\end{array}
$$

hence $(\sigma a, \sigma N|\sigma K) = \sigma(a, N|K)\sigma^{-1}$ for $a \in A_K$, where for $\sigma \in G$, the maps $A_K \xrightarrow{\sigma} A_{\sigma K}$ and $G^{\mathrm{ab}}_{N|K} \xrightarrow{\sigma^*} G^{\mathrm{ab}}_{\sigma N|\sigma K}$ are $a \mapsto \sigma a$ and $\tau \mapsto \sigma\tau\sigma^{-1}$.

All statements follow essentially from the formulas for the behavior of fundamental classes under extension given in Proposition (1.6). More precisely:

a) Let $\chi \in \chi(G_{L|K}) = H^1(G_{L|K}, \mathbb{Q}/\mathbb{Z})$, $\inf \chi \in H^1(G_{N|K}, \mathbb{Q}/\mathbb{Z})$. Then

$$
\begin{aligned}
\chi(\pi(a, N|K)) &= \inf \chi((a, N|K)) = \mathrm{inv}_{N|K}(\overline{a} \cup \delta(\inf \chi)) = \mathrm{inv}_{N|K}(\overline{a} \cup \inf(\delta\chi)) \\
&= \mathrm{inv}_{N|K}(\inf(\overline{a} \cup (\delta\chi))) = \mathrm{inv}_{L|K}(\overline{a} \cup \delta\chi) = \chi(a, L|K)
\end{aligned}
$$

by (1.10). Since this identity holds for all characters $\chi \in \chi(G_{L|K})$, we obtain $\pi(a, N|K) = (a, L|K)$.

For the proof of b) and c) we need to convince ourselves that the diagrams

$$
\begin{array}{ccccccc}
A_K & \longrightarrow & H^0(N|K) & \xleftarrow[\sim]{u_{N|K} \cup} & H^{-2}(G_{N|K}, \mathbb{Z}) & \xrightarrow{\sim} & G^{\mathrm{ab}}_{N|K} \\
{\scriptstyle \mathrm{incl}} \Big\downarrow & & {\scriptstyle \mathrm{res}} \Big\downarrow & & {\scriptstyle \mathrm{res}} \Big\downarrow & & \Big\downarrow {\scriptstyle \mathrm{Ver}} \\
A_L & \longrightarrow & H^0(N|L) & \xleftarrow[\sim]{u_{N|L} \cup} & H^{-2}(G_{N|L}, \mathbb{Z}) & \xrightarrow{\sim} & G^{\mathrm{ab}}_{N|L},
\end{array}
$$

$$
\begin{array}{ccccccc}
A_L & \longrightarrow & H^0(N|L) & \xleftarrow[\sim]{u_{N|L} \cup} & H^{-2}(G_{N|L}, \mathbb{Z}) & \xrightarrow{\sim} & G^{\mathrm{ab}}_{N|L} \\
{\scriptstyle N_{L|K}} \Big\downarrow & & {\scriptstyle \mathrm{cor}} \Big\downarrow & & {\scriptstyle \mathrm{cor}} \Big\downarrow & & \Big\downarrow {\scriptstyle \kappa} \\
A_K & \longrightarrow & H^0(N|K) & \xleftarrow[\sim]{u_{N|K} \cup} & H^{-2}(G_{N|K}, \mathbb{Z}) & \xrightarrow{\sim} & G^{\mathrm{ab}}_{N|K}.
\end{array}
$$

commute. This follows for the left squares from I, (4.9) and from the definition given in I, p. 38. For the right squares we refer to I, (4.10) and I,

(4.13). The middle squares commute since for $z \in H^{-2}(G_{N|K}, \mathbb{Z})$, resp. $z' \in H^{-2}(G_{N|L}, \mathbb{Z})$, we have by (1.6) and I, (5.4) the following identities

$$\mathrm{res}(u_{N|K} \cup z) = \mathrm{res}(u_{N|K}) \cup (\mathrm{res}\, z) = u_{N|L} \cup (\mathrm{res}\, z), \quad \text{resp.}$$
$$\mathrm{cor}(u_{N|L} \cup z') = \mathrm{cor}(\mathrm{res}(u_{N|K}) \cup z') = u_{N|K} \cup (\mathrm{cor}\, z').$$

The proof of d) is analogous and left to the reader.

Note that we could not argue similarly in the proof of (a), since inflation is only defined for positive dimensions; here Formula (1.10) has proved useful.

The essential statements of class field theory are the reciprocity law (1.9) and the properties of the norm residue symbol. In a concrete case, for example, for local or global class field theory[4], the development of the theory will be made considerably more concise by the abstract presentation given in this section, that is, by anticipating the purely group-theoretic part. It leaves us the task to verify Axioms I and II of class formations; however, we admit that there are only very few cases where this is easy.

We want to extract some consequences of the theorems we have proved so far.

If $L|K$ is normal, then the abelianization $G_{L|K}^{\mathrm{ab}}$ is the Galois group of the maximal abelian extension $L^{\mathrm{ab}}|K$ contained in L, and the reciprocity law gives an isomorphism between the Galois group of this extension and the norm residue group $A_K/N_{L|K}A_L$. We will now show that these abelian extensions are uniquely determined by their norm groups, in fact, that the entire structure of these abelian extensions of K is uniquely reflected in the group A_K of the given base field K.

A subgroup I of A_K is called a **norm group** if there is a normal field extension $L|K$ such that $I = N_{L|K}A_L$. The following theorem shows that every norm group $I = N_{L|K}A_L$ is indeed the norm group of an abelian extension of K, namely, the norm group of the maximal abelian field L^{ab} in L.

(1.12) Theorem. *Let $L|K$ be a normal extension and L^{ab} the maximal abelian extension of K contained in L. Then*

$$N_{L|K}A_L = N_{L^{\mathrm{ab}}|K}A_{L^{\mathrm{ab}}} \subseteq A_K.$$

---

[4] In addition to these, there are other interesting examples of class formations. For example, the theory of Kummer fields (cf. Part III, §1). One can even show that for each profinite group G there is a G-module A such that (G, A) is a class formation.

Proof. The inclusion $N_{L|K}A_L \subseteq N_{L^{ab}|K}A_{L^{ab}}$ follows from the multiplicativity of the norm. The reciprocity law gives isomorphisms

$$A_K/N_{L|K}A_L \cong G_{L|K}^{ab} = G_{L^{ab}|K} \cong A_K/N_{L^{ab}|K}A_{L^{ab}},$$

and $(A_K : N_{L|K}A_L) = (A_K : N_{L^{ab}|K}A_{L^{ab}}) < \infty$ implies that we have the equality $N_{L|K}A_L = N_{L^{ab}|K}A_{L^{ab}}$.

(1.13) Corollary. *The index* $(A_K : N_{L|K}A_L)$ *divides the degree* $[L : K]$ *with equality if and only if* $L|K$ *is abelian.*

In fact, $(A_K : N_{L|K}A_L) = (A_K : N_{L^{ab}|K}A_{L^{ab}}) = [L^{ab} : K]$ is a divisor of $[L : K]$, and is equal to $[L : K]$ if and only if $L = L^{ab}$, i.e., if and only if $L|K$ is abelian.

(1.14) Theorem. *The norm groups* I *of* A_K *form a lattice. The map*

$$L \longmapsto I_L = N_{L|K}A_L$$

gives an inclusion reversing isomorphism between the lattice of abelian extensions L *of* K *and the lattice of norm groups* I *of* A_K. *Hence we have*

$$I_{L_1} \supseteq I_{L_2} \Longleftrightarrow L_1 \subseteq L_2; \quad I_{L_1 \cdot L_2} = I_{L_1} \cap I_{L_2}; \quad I_{L_1 \cap L_2} = I_{L_1} \cdot I_{L_2},$$

if L_1 *and* L_2 *are abelian extension fields.*

Moreover, every group $I \subseteq A_K$ *containing a norm group is itself a norm group.*

Proof. If L_1 and L_2 are two abelian extensions of K, it follows from the multiplicativity of the norm that $I_{L_1 \cdot L_2} \subseteq I_{L_1} \cap I_{L_2}$. If $a \in I_{L_1} \cap I_{L_2}$, then the element $(a, L_1 \cdot L_2|K)$ has trivial projections $(a, L_1|K) = 1$ and $(a, L_2|K) = 1$ in $G_{L_1|K}$ and $G_{L_2|K}$, i.e., $(a, L_1 \cdot L_2|K) = 1$, so that $a \in I_{L_1 \cdot L_2}$. This proves $I_{L_1 \cdot L_2} = I_{L_1} \cap I_{L_2}$. Given this, we obtain further

$$I_{L_1} \supseteq I_{L_2} \Leftrightarrow I_{L_1} \cap I_{L_2} = I_{L_2} = I_{L_1 \cdot L_2} \Leftrightarrow [L_1 \cdot L_2 : K] = [L_2 : K] \Leftrightarrow L_1 \subseteq L_2.$$

Since every norm group is already the norm group of an abelian extension ((1.12)), we conclude that the map $L \mapsto I_L$ is a inclusion reversing bijection between the set of abelian extensions $L|K$ and the set of norm groups. The remaining statements are obvious consequences of this correspondence.

The above theorem shows that to understand the abelian extensions it is important to give a characterization of the norm groups using only intrinsic properties of the group A_K of the underlying base field K. For the concrete class formations we are interested in such a characterization is possible, because in these cases there is a canonical topology on the group A_K and the norm groups turn out to be the closed subgroups of finite index. This result is also called the **Existence Theorem**, since it shows the existence of an abelian extension L which has a given closed subgroup I of finite index in

A_K as its norm group. This (uniquely determined) field is called the **class field** associated with the group I. One can derive such an existence theorem in the theory of abstract class formations by adding certain existence axioms to Axioms I and II. However, since we do not need this for our applications, we only refer the interested reader to [42].

To complete this section we consider the norm residue symbol from a universal viewpoint. If we start with a field K, then the groups $G_{L|K}^{\mathrm{ab}}$ form a projective system of groups, namely, the projective system of Galois groups of all abelian extensions of K. We denote the projective limit of this system by

$$G_K^{\mathrm{ab}} = \varprojlim G_{L|K}^{\mathrm{ab}}$$

In the case where we are dealing with actual field extensions $L|K$, G_K^{ab} is the Galois group of the maximal abelian extension over K. We can also write

$$G_K^{\mathrm{ab}} = \varprojlim G_{L|K},$$

where L ranges over all abelian extensions of K. For every $a \in A_K$ the elements $(a, L|K) \in G_{L|K}^{\mathrm{ab}}$ form by (1.11a) a compatible system of elements in the projective system of $G_{L|K}^{\mathrm{ab}}$. Taking the limit, we obtain a unique element

$$(a, K) = \varprojlim(a, L|K) \in G_K^{\mathrm{ab}}$$

which is called the **universal norm residue symbol** of K. If

$$\pi_L : G_K^{\mathrm{ab}} \longrightarrow G_{L|K}^{\mathrm{ab}}$$

are the individual projections from G_K^{ab} onto the Galois group $G_{L|K}^{\mathrm{ab}}$, then the element $(a, K) \in G_K^{\mathrm{ab}}$ is uniquely determined by the identities

$$\pi_L(a, K) = (a, L|K).$$

The universal norm residue symbol yields a homomorphism $A_K \to G_K^{\mathrm{ab}}$ whose kernel and image we describe in the following theorem:

(1.15) Theorem. *The kernel of the homomorphism*

$$A_K \xrightarrow{(\ ,K)} G_K^{\mathrm{ab}}$$

is equal to the intersection of all norm groups

$$D_K = \bigcap_L N_{L|K} A_L,$$

and its image is dense in G_K^{ab} (with respect to the normal-subgroup topology).

Proof. We have $(a, K) = \varprojlim(a, L|K) = 1$ if and only if $(a, L|K) = 1$ for all normal extensions $L|K$, therefore if and only if $a \in D_K = \bigcap N_{L|K} A_L$. The density of the image follows equally easily: If $\sigma \in G_K^{\mathrm{ab}}$, then the sets $\sigma \cdot H$ form a fundamental system of neighborhoods of σ, where H runs through all the open subgroups of G_K^{ab}. But if H is an open subgroup, then $G_K^{\mathrm{ab}}/H =$

$G_{L|K}$ is the Galois group of an abelian extension $L|K$, and since the norm residue symbol $(\, , L|K) : A_K \to G_{L|K}$ is surjective, we find an $a \in A_K$ with $\pi_L(a, K) = (a, L|K) = \pi_L \sigma$, i.e., $(a, K) \in \sigma \cdot H$.

§ 2. Galois Cohomology

Let $L|K$ be a finite Galois field extension and $G = G_{L|K}$ its Galois group. Given such a field extension L, we immediately have two natural G-modules, namely, the additive L^+ and the multiplicative group $L^\times$ of L. The additive group is cohomologically uninteresting, because of the following result:

(2.1) Theorem. $H^q(G, L^+) = 0$ for all q.

This follows from the existence of a normal basis of $L|K$. In fact, if $c \in L$ is chosen in such a way that $\{\sigma c \mid \sigma \in G\}$ is a basis of $L|K$, then $L^+ = \bigoplus_{\sigma \in G} K^+ \cdot \sigma c = \bigoplus_{\sigma \in G} \sigma(K^+ \cdot c)$, which means that L^+ is a G-induced module. Therefore, by I, (3.13) all of its cohomology groups are trivial.

On the other hand, for the multiplicative group $L^\times$ we have the following, very important theorem:

(2.2) Theorem (Hilbert-Noether). $H^1(G, L^\times) = 1$.

Proof. Let $a_\sigma \in L^\times$ be a 1-cocycle of the G-module $L^\times$. If $c \in L^\times$, consider

$$b = \sum_{\sigma \in G} a_\sigma \cdot \sigma c.$$

Since the automorphisms σ are linearly independent (cf. [7], Ch. V, §7, n° 5), there is an element $c \in L^\times$ such that $b \neq 0$. Therefore

$$\tau(b) = \sum_{\sigma \in G} \tau a_\sigma(\tau \sigma c) = \sum_{\sigma \in G} a_\tau^{-1} \cdot a_{\tau\sigma}(\tau\sigma c) = a_\tau^{-1} \cdot b,$$

i.e., $a_\tau = \tau(b^{-1})/b^{-1}$. Hence a_τ is a 1-coboundary.

Theorem (2.2) is a generalization of the well-known "Hilbert's Theorem 90":

(2.3) Theorem (Hilbert). Let $L|K$ be a cyclic extension, and let σ be a generator of G. If $x \in L^\times$ with $N_{L|K}x = 1$, then there is a $c \in L^\times$ such that

$$x = \frac{\sigma c}{c}.$$

This theorem is just a reformulation of $N_{L|K} L^\times = (L^\times)^{\sigma-1}$, and therefore of $H^{-1}(G, L^\times) \cong H^1(G, L^\times) = 1$.

Theorem (2.2) says that the finite Galois extensions $L|K$ of K constitute a field formation in the sense of §1. In such a formation we can think of the cohomology groups $H^2(L|K)$ as the elements of the union

$$Br(K) = H^2(\ \ |K) = \bigcup_L H^2(L|K)$$

by viewing the inflation maps (which are injective because $H^1(L|K) = 1$) as the inclusions. $Br(K)$ is also called the **Brauer group** of the field K. It is an abstract variant of the well-known Brauer group in the theory of algebras, which arises as follows: Consider all central simple algebras over K. By Wedderburn's Theorem every such algebra A is isomorphic to a full matrix algebra $M_n(D)$ over a division ring $D|K$, which is up to isomorphism uniquely determined by A. Two algebras are in the same class if the corresponding division rings are isomorphic. The tensor product of two central simple algebras is again a central simple algebra, and induces a multiplication on the set of algebra classes, which makes this set a group, namely the Brauer group. We briefly describe how this group is obtained from $Br(K) = H^2(\ \ |K)$:

Let $\bar{c} \in Br(K) = H^2(\ \ |K)$, say $\bar{c} \in H^2(L|K)$, and let c be a 2-cocycle of the class $\bar{c}$. With each $\sigma \in G_{L|K}$ we associate a basis element u_σ and form the K-vector space $A = \bigoplus_{\sigma \in G_{L|K}} L \cdot u_\sigma$. In this vector space, the formulas

$$u_\sigma \cdot \lambda = (\sigma\lambda) \cdot u_\sigma \quad (\lambda \in L), \qquad u_\sigma \cdot u_\tau = c(\sigma, \tau) \cdot u_{\sigma\tau},$$

define a multiplication that makes A into a central simple algebra over K. Another 2-cocycle of $\bar{c}$ yields an equivalent algebra, and this construction gives an isomorphism between the group $H^2(\ \ |K)$ and the Brauer group of algebras (cf. [1]).

Before the introduction of cohomology theory, algebras were used to describe local class field theory (cf., e.g., [38]); we remark that the use of cohomology has led to considerable simplifications.

For finite fields we can derive the following consequence from Theorem (2.2):

(2.4) Corollary. If $L|K$ is an extension of a finite field, then

$$H^q(G_{L|K}, L^\times) = 1 \quad \text{for all } q.$$

Proof. The group $G_{L|K}$ is cyclic. Since $L^\times$ is a finite $G_{L|K}$-module, we have for the Herbrand quotient $h(L^\times) = |H^0(G_{L|K}, L^\times)|/|H^1(G_{L|K}, L^\times)| = 1$. Hence $H^q(G_{L|K}, L^\times) = 1$ for $q = 0, 1$, and therefore for all q.

§ 3. The Multiplicative Group of a p-adic Number Field

Let K be a p-adic number field, that is, a complete discrete valuation field of characteristic 0 with finite residue field[5]. We introduce the following notation. Let v be a discrete valuation of K, which we always think of as normalized so that its smallest positive value is 1,

$\mathcal{o} = \{x \in K \mid v(x) \geq 0\}$ the valuation ring,

$\mathfrak{p} = \{x \in K \mid v(x) > 0\}$ the maximal ideal,

$\overline{K} = \mathcal{o}/\mathfrak{p}$ the residue field of K, p the characteristic of $\overline{K}$,

$U = \mathcal{o} \smallsetminus \mathfrak{p}$ the unit group,

$U^1 = 1 + \mathfrak{p}$ the group of principal units, and

$U^n = 1 + \mathfrak{p}^n$ the higher unit groups.

We denote by q the number of elements in the residue field $\overline{K}$, thus $q = (\mathcal{o} : \mathfrak{p})$. If f is the degree of $\overline{K}$ over the prime field of p elements, then $q = p^f$.

Aside from v we also consider the normalized multiplicative absolute value $| \ |_\mathfrak{p}$, which arises as follows: If $x \in \mathcal{o}$, consider the absolute norm of x,

$$\mathfrak{N}(x) = (\mathcal{o} : x\mathcal{o}) = (\mathcal{o} : \mathfrak{p})^{v(x)} = p^{f \cdot v(x)},$$

and set
$$|x|_\mathfrak{p} = \mathfrak{N}(x)^{-1}.$$

If $x \in K \smallsetminus \mathcal{o}$, then $x^{-1} \in \mathcal{o}$, and we define $|x|_\mathfrak{p} = |x^{-1}|_\mathfrak{p}^{-1} = \mathfrak{N}(x^{-1})$.

(3.1) Proposition. *The group $K^\times$ has the direct decomposition*
$$K^\times = U \times (\pi),$$
where π is a prime element of $\mathfrak{p}$ and $(\pi) = \{\pi^k\}_{k \in \mathbb{Z}}$ is the infinite cyclic subgroup of $K^\times$ generated by π.

This is clear since with respect to a fixed prime element π, every $x \in K^\times$ has a unique decomposition $x = u \cdot \pi^k$, $u \in U$. Thus the short exact sequence
$$1 \longrightarrow U \longrightarrow K^\times \xrightarrow{v} \mathbb{Z} \longrightarrow 0$$

---

[5] We do not use the completeness of the field K. We only need that the valuation is absolutely indecomposable. The local class field theory developed here is therefore valid without restrictions and verbatim for fields of characteristic 0 with a henselian valuation, i.e., an absolutely indecomposable discrete valuation v with finite residue field. Only in §7 one needs some modifications.

splits, and has the group $(\pi) \cong \mathbb{Z}$ as its group of representatives. We remark that since the choice of the prime element $\pi \in \mathfrak{p}$ is arbitrary, there is no distinguished group of representatives.

In the unit group U consider the decreasing chain of higher unit groups U^n:

$$U \supset U^1 \supset U^2 \supset U^3 \supset \cdots .$$

The following result shows that the factor groups of this chain are finite:

(3.2) Proposition. $U/U^1 \cong \bar{K}^\times$ and $U^n/U^{n+1} \cong \bar{K}^+$ for $n \geq 1$.

Proof. The map that takes each $u \in U$ to its residue class $u \bmod \mathfrak{p} \in \bar{K}^\times$ defines a homomorphism from U onto $\bar{K}^\times$ with the kernel U^1. To show $U^n/U^{n+1} \cong \bar{K}^+$, we choose a prime element π. It is easy to see that the map $1 + a \cdot \pi^n \mapsto a \bmod \mathfrak{p}$ defines a homomorphism from U^n onto the additive group $\bar{K}^+$ with kernel U^{n+1}.

(3.3) Proposition. *The unit group U is an open and closed compact subgroup of $K^\times$ with respect to the valuation topology*[6].

Proof. If $u \in U$, then $\{x \in K^\times \mid v(x - u) > 0\} = u + \mathfrak{p}$ is evidently a neighborhood of u, which is entirely contained in U. Hence U is open in $K^\times$. The complement of U is the union of (open) cosets of U, therefore U is also closed.

Let $\mathfrak{S}$ be a system of open subgroups of $K^\times$ which covers U. Assume that U cannot be covered by finitely many sets from $\mathfrak{S}$. Then the same is true for a coset $u_1 \cdot U^1 \subseteq U$, since the index $(U : U^1)$ is finite. In $u_1 \cdot U^1$ there exists again among the finitely many cosets $u_2 \cdot U^2 \subseteq u_1 \cdot U^1$ one which cannot be covered by finitely many sets from $\mathfrak{S}$. Continuing in this way, we obtain a chain

$$u_1 \cdot U^1 \supseteq u_2 \cdot U^2 \supseteq u_3 \cdot U^3 \supseteq \cdots ,$$

and since U is complete as a closed subgroup of $K^\times$, there is a unit $u \in U$ such that $u \cdot U^n = u_n \cdot U^n$ for $n = 1, 2, \ldots$ The $u \cdot U^n = u + \mathfrak{p}^n$ form an open nested sequence of neighborhoods of u, and if S is a set in $\mathfrak{S}$ containing u, there is an n with $u \cdot U^n = u_n \cdot U^n \subseteq S$, which is a contradiction. Hence U is compact[7].

(3.4) Corollary. *The group $K^\times$ is locally compact*[6].

---

[6] For a henselian field in the sense of [5] we have to substitute for compact, resp. in (3.4) locally compact, relative compact resp. relative locally compact.

[7] One can also argue as follows: the groups $U^n = 1 + \mathfrak{p}^n$ form a fundamental system of neighborhoods of the unit element $1 \in U$. Therefore $U = \varprojlim U^n/U^{n+1}$ ($U^0 = U$), where the inverse limit $\varprojlim U^n/U^{n+1}$ is compact as a profinite group.

If $x \in K^\times$, then $x \cdot U$ is an open and compact neighborhood of x by (3.3).

(3.5) Lemma. *If m is a positive integer, then the map $x \mapsto x^m$ yields for sufficiently large n an isomorphism*
$$U^n \longrightarrow U^{n+v(m)}.$$

Proof. If π is a prime element of $\mathfrak{p}$ and $x = 1 + a \cdot \pi^n \in U^n$, then
$$x^m = 1 + m \cdot a \cdot \pi^n + \binom{m}{2} a^2 \cdot \pi^{2n} + \cdots \equiv 1 \bmod \mathfrak{p}^{n+v(m)},$$
and therefore $x^m \in U^{n+v(m)}$ for sufficiently large n.

To prove this map is surjective we have to show that for every $a \in \mathcal{o}$ there exists an element $x \in \mathcal{o}$ such that
$$1 + a \cdot \pi^{n+v(m)} = (1 + x \cdot \pi^n)^m,$$
i.e., $1 + a \cdot \pi^{n+v(m)} = 1 + m \cdot \pi^n \cdot x + \pi^{2n} \cdot f(x)$, where $f(x)$ is an integral polynomial in x. Obviously $m = u \cdot \pi^{v(m)}$, $u \in U$, and we obtain the equation
$$-a + u \cdot x + \pi^{n-v(m)} \cdot f(x) = 0.$$
If $n > v(m)$, then Hensel's Lemma clearly gives a solution $x \in \mathcal{o}$.

If additionally n is chosen so large that U^n contains no m-th root of unity, then our map is also injective. ∎

(3.6) Corollary. *If m is a positive integer, then the group of m-th powers $(K^\times)^m$ is an open subgroup of $K^\times$. Furthermore,*
$$\bigcap_{m=1}^{\infty} (K^\times)^m = 1.$$

Proof. If $x^m \in (K^\times)^m$, then for a sufficiently large positive integer n the set
$$x^m \cdot U^{n+v(m)} = (x \cdot U^n)^m \subseteq (K^\times)^m$$
is an open neighborhood of x^m. If $a \in \bigcap_{m=1}^{\infty} (K^\times)^m$, then trivially $a \in U$, and therefore $a \in \bigcap_{m=1}^{\infty}(U)^m$, i.e., $a = u_m^m$, $u_m \in U$, for all m. Now if n is an arbitrary positive integer and $m = (U : U^n)$, then $a = u_m^m \in U^n$, and hence $a \in \bigcap_{n=1}^{\infty} U^n = \{1\}$. ∎

We denote by $\mu_m(K)$ the group of m-th roots of unity in K and prove the

(3.7) Proposition. *The group $(K^\times)^m$ has finite index in $K^\times$; more precisely,*
$$(K^\times : (K^\times)^m) = m \cdot q^{v(m)} \cdot |\mu_m(K)| = m \cdot |m|_{\mathfrak{p}}^{-1} \cdot |\mu_m(K)|,$$
where q is the number of elements in $\overline{K}$ and $|\mu_m(K)|$ is the number of elements in $\mu_m(K)$.

For the proof we make use of the Herbrand quotient with respect to the two endomorphisms 0 and m (cf. I, p. 54). Then we have

$$(K^\times : (K^\times)^m) = q_{0,m}(K^\times) \cdot |\mu_m(K)|.$$

Since $q_{0,m}$ is multiplicative, we can decompose the first factor on the right as

$$q_{0,m}(K^\times) = q_{0,m}(K^\times/U) \cdot q_{0,m}(U/U^n) \cdot q_{0,m}(U^n).$$

Here $q_{0,m}(K^\times/U) = q_{0,m}(\mathbb{Z}) = m$ because of (3.1), $q_{0,m}(U/U^n) = 1$ since U/U^n is finite by (3.2); and $q_{0,m}(U^n) = (U^n : U^{n+v(m)}) = q^{v(m)}$ for sufficiently large n by (3.5), using the fact that $(U^i : U^{i+1}) = q$.

Putting all this together proves the above formula. At the same time we get

(3.8) Corollary. $(U : (U)^m) = |m|_{\mathfrak{p}}^{-1} \cdot |\mu_m(K)|$ [8)].

§ 4. The Class Formation of Unramified Extensions

A relatively simple example of a class formation is given by the unramified extensions of a $\mathfrak{p}$-adic number field K. Although this class formation is a special case of the more general formations we consider in the next section, we need to look at it separately first. This, because its reciprocity law is remarkably simple, and because the results obtained for this special case will be applied in the proof of the general local reciprocity law.

In what follows we consider finite extensions $L|K$ of $\mathfrak{p}$-adic number fields and append to the notation v, $\mathcal{O}$, $\mathfrak{p}$, etc. introduced in §3 the relevant field as an index, thus writing v_K, $\mathcal{O}_K$, $\mathfrak{p}_K$; v_L, $\mathcal{O}_L$, $\mathfrak{p}_L$, etc. The valuation v_K has a unique extension to L, namely the valuation $\frac{1}{e} \cdot v_L$, where e is the ramification index of $L|K$.

The extension $L|K$ is unramified when $e = 1$, i.e., if a prime element $\pi \in K$ for $\mathfrak{p}_K$ is also a prime element for $\mathfrak{p}_L$. This is equivalent to the statement that the degree of the field extension $[L : K]$ is the same as the degree $[\bar{L} : \bar{K}]$ of the residue field extension $\bar{L}|\bar{K}$.

An unramified extension $L|K$ is normal, and there is a canonical isomorphism

$$G_{L|K} \cong G_{\bar{L}|\bar{K}}$$

between the Galois group $G_{L|K}$ of the extension $L|K$ and the Galois group $G_{\bar{L}|\bar{K}}$ of the residue field extension $\bar{L}|\bar{K}$. In fact, if $\sigma \in G_{L|K}$, we obtain from

$$\bar{\sigma}(x + \mathfrak{p}_L) = \sigma x + \mathfrak{p}_L, \quad x \in \mathcal{O}_L,$$

a $\bar{K}$-automorphism $\bar{\sigma}$ of $\bar{L}$.

[8)] We denote by $(U)^m$ the group of m-th powers of U (in contrast to U^m).

The group $G_{\bar{L}|\bar{K}}$ is cyclic as the Galois group of a finite field $\bar{L}$. As one can immediately verify, it has the distinguished generating automorphism

$$\bar{x} \longmapsto \bar{x}^{q_K}, \quad \bar{x} \in \bar{L},$$

where q_K is the number of elements contained in $\bar{K}$. Because $G_{L|K} \cong G_{\bar{L}|\bar{K}}$, we also obtain a canonical K-automorphism of L which generates $G_{L|K}$.

(4.1) Definition. *The automorphism $\varphi_{L|K} \in G_{L|K}$ which is induced by the automorphism*

$$\bar{x} \longmapsto \bar{x}^{q_K}, \quad \bar{x} \in \bar{L},$$

*of the residue field $\bar{L}$ is called the **Frobenius automorphism** of $L|K$.*

(4.2) Proposition. *Let $N \supseteq L \supseteq K$ be unramified extensions of K. Then*

$$\varphi_{L|K} = \varphi_{N|K}|_L = \varphi_{N|K} G_{N|L} \in G_{L|K} \quad \text{and} \quad \varphi_{N|L} = \varphi_{N|K}^{[L:K]}.$$

Proof. This follows easily from the fact that for all $x \in \mathcal{O}_L$ we have

$$(\varphi_{L|K} x) \bmod \mathfrak{p}_L = x^{q_K} \bmod \mathfrak{p}_L = x^{q_K} \bmod \mathfrak{p}_N = (\varphi_{N|K} x) \bmod \mathfrak{p}_N,$$

and for all $x \in \mathcal{O}_N$,

$$(\varphi_{N|L} x) \bmod \mathfrak{p}_N = x^{q_L} \bmod \mathfrak{p}_N = x^{q_K^{[L:K]}} \bmod \mathfrak{p}_N = (\varphi_{N|K}^{[L:K]} x) \bmod \mathfrak{p}_N.$$

Because of its canonical nature and its good properties with respect to restrictions and extensions given by (4.2), the Frobenius automorphism plays a significant special role in class field theory.

The following theorem is particularly important in both local and global class field theory.

(4.3) Theorem. *Let $L|K$ be an unramified extension. Then*

$$H^q(G_{L|K}, U_L) = 1 \quad \text{for all } q.$$

Proof[9]. If we identify the group $G_{\bar{L}|\bar{K}}$ with the group $G_{L|K}$, then

$$1 \longrightarrow U_L^1 \longrightarrow U_L \longrightarrow \bar{L}^\times \longrightarrow 1$$

is an exact sequence of $G_{L|K}$-modules. Since $H^q(G_{L|K}, \bar{L}^\times) = 1$ (cf. (2.4)), it follows that $H^q(G_{L|K}, U_L) \cong H^q(G_{L|K}, U_L^1)$.

A prime element $\pi \in K$ for $\mathfrak{p}_K$ is also a prime element of $\mathfrak{p}_L$. Thus the map

$$U_L^{n-1} \to \bar{L}^+, \; 1 + a \cdot \pi^{n-1} \longmapsto a \quad \bmod \mathfrak{p}_L, \quad a \in \mathcal{O}_L,$$

---

[9] For the proof of this theorem one uses usually the completeness of the field L (cf. [42]). We avoid this, so that our entire exposition also holds verbatim for henselian fields in the sense of [5].

defines a $G_{L|K}$-homomorphism, and from the exact sequence of $G_{L|K}$-modules

$$1 \longrightarrow U_L^n \longrightarrow U_L^{n-1} \longrightarrow \bar{L}^+ \longrightarrow 0$$

we obtain, using that $H^q(G_{L|K}, \bar{L}^+) = 0$ for all q by (2.1), the isomorphism

$$H^q(G_{L|K}, U_L^n) \cong H^q(G_{L|K}, U_L^{n-1}).$$

Thus it follows that the injection $U_L^n \to U_L$ induces an isomorphism

$$H^q(G_{L|K}, U_L^n) \longrightarrow H^q(G_{L|K}, U_L).$$

If m is a positive integer, the map $x \mapsto x^m$ defines a homomorphism $U_L \xrightarrow{m} U_L$, and by (3.5) an isomorphism $U_L^n \to U_L^{n+v(m)}$, provided n is sufficiently large. Hence we have a homomorphism $H^q(G_{L|K}, U_L) \xrightarrow{m} H^q(G_{L|K}, U_L)$, and an isomorphism $H^q(G_{L|K}, U_L^n) \xrightarrow{m} H^q(G_{L|K}, U_L^{n+v(m)})$. Consider the diagram

$$
\begin{array}{ccc}
H^q(G_{L|K}, U_L^n) & \longrightarrow & H^q(G_{L|K}, U_L) \\
{\scriptstyle m}\downarrow & & {\scriptstyle m}\downarrow \\
H^q(G_{L|K}, U_L^{n+v(m)}) & \longrightarrow & H^q(G_{L|K}, U_L)
\end{array}
$$

This diagram obviously commutes, and all maps except for the right vertical map are known to be bijections. Hence it follows that the homomorphism

$$H^q(G_{L|K}, U_L) \xrightarrow{m} H^q(G_{L|K}, U_L),$$

that sends every cohomology class c to its m-th power c^m is also a bijection for all m. But the elements of $H^q(G_{L|K}, U_L)$ have finite order (cf. I, (3.16)), so that we must have $H^q(G_{L|K}, U_L) = 1$.

For $q = 0$ we obtain the

(4.4) Corollary. *If $L|K$ is unramified, then*

$$U_K = N_{L|K} U_L.$$

Hence in the unramified case every unit is also a norm.

We now show that the unramified extensions $L|K$ form a class formation with respect to the multiplicative group $L^\times$. To do this, we have to specify an invariant map satisfying Axiom II (cf. §1, (1.3)). We proceed as follows. From the long exact cohomology sequence associated with the sequence

$$1 \longrightarrow U_L \longrightarrow L^\times \xrightarrow{v_L} \mathbb{Z} \longrightarrow 0,$$

we obtain, using that $H^q(G_{L|K}, U_L) = 1$, the isomorphism

$$H^2(G_{L|K}, L^\times) \xrightarrow{\bar{v}} H^2(G_{L|K}, \mathbb{Z}).$$

Moreover, the exact sequence $0 \to \mathbb{Z} \to \mathbb{Q} \to \mathbb{Q}/\mathbb{Z} \to 0$, together with the fact that $\mathbb{Q}$ is cohomologically trivial, implies that the connecting map

$$H^2(G_{L|K}, \mathbb{Z}) \xrightarrow{\delta^{-1}} H^1(G_{L|K}, \mathbb{Q}/\mathbb{Z}) = \mathrm{Hom}(G_{L|K}, \mathbb{Q}/\mathbb{Z}) = \chi(G_{L|K})$$

is an isomorphism. If $\chi \in \chi(G_{L|K})$, then $\chi(\varphi_{L|K}) \in \frac{1}{[L:K]}\mathbb{Z}/\mathbb{Z} \subseteq \mathbb{Q}/\mathbb{Z}$, and since the Frobenius automorphism $\varphi_{L|K}$ generates the group $G_{L|K}$, the map

$$H^1(G_{L|K}, \mathbb{Q}/\mathbb{Z}) = \chi(G_{L|K}) \xrightarrow{\varphi} \frac{1}{[L:K]}\mathbb{Z}/\mathbb{Z}$$

is also an isomorphism. Taking the composition of these three isomorphisms,

$$H^2(G_{L|K}, L^\times) \xrightarrow{\bar{v}} H^2(G_{L|K}, \mathbb{Z}) \xrightarrow{\delta^{-1}} H^1(G_{L|K}, \mathbb{Q}/\mathbb{Z}) \xrightarrow{\varphi} \frac{1}{[L:K]}\mathbb{Z}/\mathbb{Z},$$

we obtain the desired map

(4.5) Definition. *If $L|K$ is an unramified extension, define*

$$\mathrm{inv}_{L|K} : H^2(G_{L|K}, L^\times) \longrightarrow \frac{1}{[L:K]}\mathbb{Z}/\mathbb{Z}$$

to be the isomorphism $\mathrm{inv}_{L|K} = \varphi \circ \delta^{-1} \circ \bar{v}$.

For simplicity, we now set

$$H^q(L|K) = H^q(G_{L|K}, L^\times).$$

Let K_0 be a fixed $\mathfrak{p}$-adic number field, and let T be the maximal unramified field extension of K_0, thus the union of all unramified extensions $L|K_0$; the field T is also called the **inertia field** over K_0. We denote by $G_{T|K_0}$ the Galois group of $T|K_0$.

(4.6) Theorem. *The formation $(G_{T|K_0}, T^\times)$ is a class formation with respect to the invariant map defined in (4.5).*

Proof. Axiom I is always satisfied by (2.2): $H^1(L|K) = 1$. For the proof of Axiom II a) and b) we need to prove that the following two diagrams commute

$$
\begin{array}{ccccccc}
H^2(L|K) & \xrightarrow{\bar{v}} & H^2(G_{L|K}, \mathbb{Z}) & \xrightarrow{\delta^{-1}} & H^1(G_{L|K}, \mathbb{Q}/\mathbb{Z}) & \xrightarrow{\varphi} & \frac{1}{[L:K]}\mathbb{Z}/\mathbb{Z} \\
\downarrow{\scriptstyle\mathrm{incl}} & & \downarrow{\scriptstyle\mathrm{inf}} & & \downarrow{\scriptstyle\mathrm{inf}} & & \downarrow{\scriptstyle\mathrm{incl}} \\
H^2(N|K) & \xrightarrow{\bar{v}} & H^2(G_{N|K}, \mathbb{Z}) & \xrightarrow{\delta^{-1}} & H^1(G_{N|K}, \mathbb{Q}/\mathbb{Z}) & \xrightarrow{\varphi} & \frac{1}{[N:K]}\mathbb{Z}/\mathbb{Z},
\end{array}
$$

$$
\begin{array}{ccccccc}
H^2(N|K) & \xrightarrow{\bar{v}} & H^2(G_{N|K}, \mathbb{Z}) & \xrightarrow{\delta^{-1}} & H^1(G_{N|K}, \mathbb{Q}/\mathbb{Z}) & \xrightarrow{\varphi} & \frac{1}{[N:K]}\mathbb{Z}/\mathbb{Z} \\
\downarrow{\scriptstyle\mathrm{res}} & & \downarrow{\scriptstyle\mathrm{res}} & & \downarrow{\scriptstyle\mathrm{res}} & & \downarrow{\scriptstyle[L:K]} \\
H^2(N|L) & \xrightarrow{\bar{v}} & H^2(G_{N|L}, \mathbb{Z}) & \xrightarrow{\delta^{-1}} & H^1(G_{N|L}, \mathbb{Q}/\mathbb{Z}) & \xrightarrow{\varphi} & \frac{1}{[N:L]}\mathbb{Z}/\mathbb{Z}.
\end{array}
$$

where $N \supseteq L \supseteq K$ are two unramified extensions of K. That the left squares commute follows immediately from the behavior of 2-cocycles under the maps

$\bar{v}$, inf, and res. The middle squares commute because the inflation and restriction maps are compatible with the connecting homomorphism δ (cf. I, (4.4) and I, (4.5)).

To prove the commutativity of the right squares, let $\chi \in H^1(G_{L|K}, \mathbb{Q}/\mathbb{Z})$ and $\chi \in H^1(G_{N|K}, \mathbb{Q}/\mathbb{Z})$ respectively. From (4.2) we have the formulas

$$\inf \chi(\varphi_{N|K}) = \chi(\varphi_{N|K}G_{N|L}) = \chi(\varphi_{L|K}), \text{ and}$$
$$\operatorname{res} \chi(\varphi_{N|L}) = \chi(\varphi_{N|L}) = \chi(\varphi_{N|K}^{[L:K]}) = [L:K] \cdot \chi(\varphi_{N|K}),$$

which complete the proof.

We could now apply the entire theory developed in §1 to this special class formation. However, we will not pursue this here, since we will do this for more general, not necessarily unramified, extensions in the next section.

If T is the maximal unramified field extension of K_0, and therefore the maximal unramified extension of every unramified extension $K|K_0$, then we set

$$H^2(T|K) = \bigcup_L H^2(L|K),$$

where $L|K$ runs through all (finite) unramified extensions of K. Here we view the inflation maps as inclusions (cf. §1, p. 66), so that for two normal extensions $N \supseteq L \supseteq K$ we have $H^2(L|K) \subseteq H^2(N|K)$. Because of the extension property of the invariant map II a), we obtain an injective homomorphism

$$\operatorname{inv}_K : H^2(T|K) \longrightarrow \mathbb{Q}/\mathbb{Z}$$

(cf. §1, p. 68). This homomorphism is even bijective, since $\mathbb{Q}/\mathbb{Z} = \bigcup_{n=1}^{\infty} \frac{1}{n}\mathbb{Z}/\mathbb{Z}$, and since for every positive integer n there exists (exactly) one unramified extension $L|K$ of degree $n = [L:K]$, for which we have the bijective homomorphism $\operatorname{inv}_{L|K} : H^2(L|K) \to \frac{1}{n}\mathbb{Z}/\mathbb{Z}$. Thus we have shown:

(4.7) Proposition. $H^2(T|K) \cong \mathbb{Q}/\mathbb{Z}$.

If $L|K$ is an unramified extension, the Galois group $G_{L|K}$ is cyclic and thus coincides with its abelianization $G_{L|K}^{\mathrm{ab}}$. Hence the norm residue symbol $(\ , L|K)$ yields the exact sequence

$$1 \longrightarrow N_{L|K}L^\times \longrightarrow K^\times \xrightarrow{(\ , L|K)} G_{L|K} \longrightarrow 1 \, .$$

What is special about the reciprocity law in the unramified case is that the norm residue symbol has a very simple, explicit description.

(4.8) Theorem. *Let $L|K$ be unramified, and $a \in K^\times$. Then*

$$(a, L|K) = \varphi_{L|K}^{v_K(a)}.$$

Proof. If $\chi \in \chi(G_{L|K})$, $\delta\chi \in H^2(G_{L|K}, \mathbb{Z})$ and $\overline{a} = a \cdot N_{L|K} L^\times \in H^0(L|K)$,

$$\chi(a, L|K) = \mathrm{inv}_{L|K}(\overline{a} \cup \delta\chi)$$

by (1.10). This formula, together with (4.5) implies that

$$\chi(a, L|K) = \mathrm{inv}_{L|K}(\overline{a} \cup \delta\chi) = \varphi \circ \delta^{-1} \circ \overline{v}(\overline{a} \cup \delta\chi) = \varphi \circ \delta^{-1}(v_K(a) \cdot \delta\chi)$$
$$= \varphi(v_K(a) \cdot \chi) = v_K(a) \cdot \chi(\varphi_{L|K}) = \chi(\varphi_{L|K}^{v_K(a)}).$$

Since this holds for all $\chi \in \chi(G_{L|K})$, it follows that $(a, L|K) = \varphi_{L|K}^{v_K(a)}$. $\quad\square$

Theorem (4.8) raises the question whether one can obtain the reciprocity law without the cohomological calculus and the notion of class formations. It appears that one could get the reciprocity law in a much more natural way by simply defining the norm residue symbol explicitly by the formula $(a, L|K) = \varphi_{L|K}^{v_K(a)}$, and then directly verifying its essential properties. In the unramified case, this is in fact possible. After closer inspection, the use of the invariant map for the present case seems to be unnecessarily complicated, in fact might appear as an attempt to actually force this idea artificially into a cohomological framework. The reason to follow this approach lies in the problem of a class field theoretical treatment of ramified field extensions. Historically, it was precisely at this point, where cohomology (in the theory of algebras) entered class field theory. In fact, for ramified extensions one cannot readily give an explicit definition of the norm residue symbol, whereas this can be done for an invariant map, which extends canonically the one constructed here to the domain of arbitrary normal extensions. We will see this in the next section.

In the class formation of unramified extensions, the extension fields L of K correspond by (1.14) precisely to the norm groups in $K^\times$. Because of (4.8), these norm groups can be given explicitly.

(4.9) Theorem. *Let K be a $\mathfrak{p}$-adic number field and π a prime element. Then*

$$U_K \times (\pi^f) \qquad {}^{10)}$$

is the norm group of the unramified extension $L|K$ of degree f.

Proof. Since $\varphi_{L|K}$ generates the group $G_{L|K}$, the degree $f = [L : K]$ is also the order of $\varphi_{L|K}$ in $G_{L|K}$. Hence an element $a \in K^\times$ lies in $N_{L|K} L^\times$ if and only if $(a, L|K) = \varphi_{L|K}^{v_K(a)} = 1$, thus if and only if $v_K(a) \equiv 0 \bmod f$, i.e., $a = u \cdot \pi^{k \cdot f}$, $k \in \mathbb{Z}$, $u \in U_K$.

---

[10] We denote by (π^f) the infinite cyclic group $\{\pi^{k \cdot f}\}_{k \in \mathbb{Z}}$ generated by the element π^f.

We end this section with some remarks about the universal norm residue symbol of our class formation (cf. §1, p. 76). Let $T|K$ be the maximal unramified extension of K. If $L|K$ ranges over all finite unramified extensions, then the projective limit

$$G_{T|K} = \varprojlim_{L} G_{L|K}$$

is the Galois group of $T|K$.

For $a \in K^{\times}$ we get the universal norm residue symbol $(a, T|K) \in G_{T|K}$ by

$$(a, T|K) = \varprojlim_{L}(a, L|K).$$

This gives a homomorphism

$$K^{\times} \xrightarrow{(\ ,T|K)} G_{T|K}.$$

If $\pi_L : G_{T|K} \to G_{L|K}$ is the canonical projection of $G_{T|K}$ onto $G_{L|K}$, then

$$\pi_L(a, T|K) = (a, L|K) = \varphi_{L|K}^{v_K(a)} \in G_{L|K}.$$

Because of (4.2), the elements $\varphi_{L|K} \in G_{L|K}$ form a compatible system of elements in the projective system of $G_{L|K}$, and we call the element

$$\varphi_K = \varprojlim_{L} \varphi_{L|K} \in G_{T|K}$$

the "universal" Frobenius automorphism of K. It has infinite order, since from $\varphi_K^n = 1$ it would immediately follow that $\pi_L(\varphi_K^n) = \varphi_{L|K}^n = 1$ for all $\varphi_{L|K}$, which is obviously impossible.

For this symbol $(\ , T|K)$, we now have the

(4.10) Theorem. *If $a \in K^{\times}$, then $(a, T|K) = \varphi_K^{v_K(a)}$. The kernel of the homomorphism*

$$K^{\times} \xrightarrow{(\ ,T|K)} G_{T|K}$$

is the unit group U_K.

Proof. If $L|K$ is an unramified extension, and $\pi_L : G_{T|K} \to G_{L|K}$ is the canonical projection from $G_{T|K}$ onto $G_{L|K}$, then

$$\pi_L(a, T|K) = (a, L|K) = \varphi_{L|K}^{v_K(a)} = \pi_L(\varphi_K^{v_K(a)}).$$

This shows that $(a, T|K) = \varphi_K^{v_K(a)}$. Hence $(a, T|K) = \varphi_K^{v_K(a)} = 1$ if and only if $v_K(a) = 0$ (φ_K has infinite order), therefore if and only if $a \in U_K$.

The class formation of unramified extensions provides an example that shows that the universal norm residue symbol is not surjective in general. In fact, its image is the infinite cyclic group generated by φ_K, which is a dense subgroup of $G_{T|K}$ isomorphic to $\mathbb{Z}$. Since this is not a profinite group, it obviously cannot coincide with $G_{T|K}$; only its completion in $G_{T|K}$ coincides with $G_{T|K}$.

§ 5. The Local Reciprocity Law

We fix a p-adic number field K_0 and let Ω denote its algebraic closure. For every normal extension $L|K$ with L finite over K_0 we again set (cf. §2, p. 78)

$$H^q(L|K) = H^q(G_{L|K}, L^\times),$$
$$Br(K) \;\; = H^2(\;\;|K) = \bigcup_{L|K} H^2(L|K) \qquad \text{(Brauer group of } K).$$

If $G = G_{\Omega|K_0}$ is the Galois group of $\Omega|K_0$, the formation $(G, \Omega^\times)$ is a field formation, because $H^1(L|K) = 1$ (cf. (2.2)). We will show in this section that it is even a class formation. For this we have to extend the invariant map introduced in §4 to ramified extensions $L|K$. The following lemma, which is also called the "second fundamental inequality", provides the key to this generalization:

(5.1) Lemma. *Let $L|K$ be a normal extension. Then the order $|H^2(L|K)|$ of $H^2(L|K)$ is a divisor of the degree $[L : K]$:*

$$|H^2(L|K)| \; \big| \; [L : K].$$

Proof. Assume first $L|K$ is cyclic of prime degree $p = [L : K]$. We show the Herbrand quotient $h(L^\times) = |H^2(L|K)|/|H^1(L|K)| = |H^2(L|K)| = p$. If $q_{0,p}$ is the Herbrand quotient with respect to the endomorphisms 0 and p, then

$$h(L^\times)^{p-1} = q_{0,p}(K^\times)^p/q_{0,p}(L^\times)$$

by I, (6.9). Using (3.7), we have for the Herbrand quotients on the right side

$$q_{0,p}(K^\times) = (K^\times : (K^\times)^p)/|K_p| = p \cdot q_K^{v_K(p)},$$
$$q_{0,p}(L^\times) = (L^\times : (L^\times)^p)/|L_p| = p \cdot q_L^{v_L(p)}.$$

If $f = [\bar{L} : \bar{K}]$ is the inertia degree and e the ramification index, then $p = e \cdot f$, $q_L = q_K^f$, and $v_L(p) = e \cdot v_K(p)$. Substitution into the above formulae yields

$$h(L^\times)^{p-1} = p^p \cdot q_K^{p \cdot v_K(p)}/p \cdot q_K^{e \cdot f \cdot v_K(p)} = p^{p-1}, \text{ i.e., } h(L^\times) = p.$$

The general case follows from this by purely cohomological methods. Since the Galois group $G_{L|K}$ is solvable, there exists a cyclic intermediate field K' over K of prime degree, $K \subseteq K' \subseteq L$. Because $H^1(L|K') = 1$, the sequence

$$1 \longrightarrow H^2(K'|K) \longrightarrow H^2(L|K) \xrightarrow{\text{res}} H^2(L|K')$$

is exact. This shows that

$$|H^2(L|K)| \; \big| \; |H^2(L|K')| \cdot |H^2(K'|K)| \,.$$

We have already shown that $|H^2(K'|K)| = [K' : K]$, and when we assume by induction on the field degree that $|H^2(L|K')| \; \big| \; [L : K']$, then it follows that

$$|H^2(L|K)| \; \big| \; [L : K'] \cdot [K' : K] = [L : K] \,.$$

The above proof makes use of the solvability of the Galois group $G_{L|K}$, which follows immediately from the fact that we have a cyclic inertia field between K and L, and above it the cyclic ramification field over which L has prime power degree. One can get around this by using I, (4.16) to reduce to the case of an extension of prime power degree, and then proceeding as above.

That we can extend the invariant map to the case of ramified extensions is obvious, once we have proved the following theorem:

(5.2) Theorem. *If $L|K$ is a normal extension and $L'|K$ is the unramified extension of the same degree $[L' : K] = [L : K]$, then*

$$H^2(L|K) = H^2(L'|K) \subseteq H^2(\quad|K).$$

Proof. It suffices to show the inclusion $H^2(L'|K) \subseteq H^2(L|K)$; if this holds, the inclusion must have an equality because $|H^2(L'|K)| = [L : K]$ (cf. (4.5)), and $|H^2(L|K)| \,\big|\, [L : K]$ by (5.1).

If $N = L \cdot L'$, and $L'|K$ is unramified, then $N|L$ is also unramified, [11]. Let $c \in H^2(L'|K) \subseteq H^2(N|K)$. It follows from the exact sequence

$$1 \longrightarrow H^2(L|K) \longrightarrow H^2(N|K) \xrightarrow{\text{res}_L} H^2(N|L),$$

that c lies in $H^2(L|K)$ if and only if $\text{res}_L c = 1$. Since $\text{res}_L c = 1$ if and only if $\text{inv}_{N|L}(\text{res}_L c) = 0$ (cf. (4.6)), our theorem follows once we have shown that

$$\text{inv}_{N|L}(\text{res}_L c) = [L : K] \cdot \text{inv}_{L'|K} c,$$

since $\text{inv}_{L'|K} c \in \frac{1}{[L:K]}\mathbb{Z}/\mathbb{Z}$. The last equality is a special case of the following lemma.

(5.3) Lemma. *Let $M|K$ be a normal extension containing the two extensions $L|K$ and $L'|K$ with $L'|K$ unramified. Then $N = L \cdot L'|L$ is also unramified [11]. If $c \in H^2(L'|K) \subseteq H^2(M|K)$, then $\text{res}_L c \in H^2(N|L) \subseteq H^2(M|L)$, and*

$$\text{inv}_{N|L}(\text{res}_L c) = [L : K] \cdot \text{inv}_{L'|K} c.$$

Proof. The fact that the 2-cocycles of the class $\text{res}_L c$ have their values in $N^\times$ (cf. §1, p. 66) implies that $\text{res}_L c \in H^2(N|L)$.

Let f be the inertia degree and e the ramification index of the (not necessarily normal) extension $L|K$. We think of the valuations v_K and v_L as extended to M. Then we have $v_L = e \cdot v_K$. By definition, the invariant map is the composite of the three isomorphisms $\bar{v}$, δ^{-1}, φ (cf. (4.5)); hence to prove the above formula it suffices to check that the following diagram commutes

---

[11] Note: If T is the maximal unramified extension of K, then $T \cdot L$ is the maximal unramified extension of L.

$$H^2(L'|K) \xrightarrow{\bar{v}_K} H^2(G_{L'|K}, \mathbb{Z}) \xrightarrow{\delta^{-1}} H^1(G_{L'|K}, \mathbb{Q}/\mathbb{Z}) \xrightarrow{\varphi} \tfrac{1}{[L':K]}\mathbb{Z}/\mathbb{Z}$$

$$\downarrow{\text{incl}} \qquad \downarrow{\text{inf}} \qquad \downarrow{\text{inf}} \qquad \downarrow{\text{incl}}$$

$$H^2(M|K) \qquad H^2(G_{M|K}, \mathbb{Z}) \qquad H^1(G_{M|K}, \mathbb{Q}/\mathbb{Z}) \qquad \tfrac{1}{[M:K]}\mathbb{Z}/\mathbb{Z}$$

$$\downarrow{\text{res}_L} \qquad \downarrow{e\cdot\text{res}} \qquad \downarrow{e\cdot\text{res}} \qquad \downarrow{\cdot[L:K]}$$

$$H^2(N|L) \xrightarrow{\bar{v}_L} H^2(G_{N|L}, \mathbb{Z}) \xrightarrow{\delta^{-1}} H^1(G_{N|L}, \mathbb{Q}/\mathbb{Z}) \xrightarrow{\varphi} \tfrac{1}{[N:L]}\mathbb{Z}/\mathbb{Z}.$$

In this diagram it is understood that the lower vertical maps only map the images of the upper vertical maps to the cohomology groups in the bottom row. That the left square commutes follows from the behavior of the 2-cocycles under the maps in question. The middle square commutes because the inflation and restriction maps commute with the homomorphism δ (cf. I, (4.4) and I, (4.5)). To see that the right square commutes, we have to consider the equation

$$\varphi_{N|L}\big|_{L'} = \varphi^f_{L'|K}$$

which is a generalization of (4.2). But this is easy to see that if $a \in L'$, then

$$\varphi_{N|L}(a) \equiv a^{q_L} \bmod \mathfrak{p}_N = a^{q^f_K} \bmod \mathfrak{p}_N = a^{q^f_K} \bmod \mathfrak{p}_{L'} = \varphi^f_{L'|K}(a).$$

Now if $\chi \in H^1(G_{L'|K}, \mathbb{Q}/\mathbb{Z})$, then

$$[L:K] \cdot \chi(\varphi_{L'|K}) = e \cdot f \cdot \chi(\varphi_{L'|K}) = e \cdot \chi(\varphi^f_{L'|K}) = e \cdot \chi(\varphi_{N|L}\big|_{L'})$$
$$= e \cdot \inf\chi(\varphi_{N|L}) = e \cdot (\text{res} \circ \inf)\chi(\varphi_{N|L}).$$

hence the right diagram commutes, which completes the proof of the lemma.

From Theorem (5.2) we have the equality

$$Br(K) = H^2(\ \ |K) = H^2(T|K) = \bigcup_{\substack{L|K \\ \text{unramified}}} H^2(L|K),$$

and because of the invariant map we obtain using (4.7) the

(5.4) Theorem. *The Brauer group $Br(K)$ of a $\mathfrak{p}$-adic number field K is canonically isomorphic to $\mathbb{Q}/\mathbb{Z}$:*

$$Br(K) \cong \mathbb{Q}/\mathbb{Z}.$$

(5.5) Definition. *Let $L|K$ be a normal extension and $L'|K$ be the unramified extension of the same degree $[L' : K] = [L : K]$, so that $H^2(L|K) = H^2(L'|K)$. Define*

$$\text{inv}_{L|K} : H^2(L|K) \longrightarrow \tfrac{1}{[L:K]}\mathbb{Z}/\mathbb{Z}$$

to be the isomorphism $\text{inv}_{L|K} c = \text{inv}_{L'|K} c$ ($c \in H^2(L|K) = H^2(L'|K)$).

With the definition of this invariant map we have reached our goal. If K_0 is a $\mathfrak{p}$-adic number field, Ω its algebraic closure and $G_{K_0} = G_{\Omega|K_0}$ is the Galois group of $\Omega|K_0$, then

(5.6) Theorem. *The formation* $(G_{K_0}, \Omega^\times)$ *is a class formation with respect to the invariant map defined in (5.5).*

Proof. Axiom I is satisfied by (2.2): $H^1(L|K) = 1$.

If $N \supseteq L \supseteq K$ are normal extensions over a fixed $\mathfrak{p}$-adic base field K_0 and $N'|K$ (resp. $L'|K$) is the unramified extensions of degree $[N' : K] = [N : K]$, (resp. $[L' : K] = [L : K]$) $(N' \supseteq L' \supseteq K)$, then for every $c \in H^2(L|K) = H^2(L'|K)$ we have, using (4.6),

$$\mathrm{inv}_{N|K} c = \mathrm{inv}_{N'|K} c = \mathrm{inv}_{L'|K} c = \mathrm{inv}_{L|K} c.$$

This proves Axiom II a). For the proof of Axiom II b), let $L|K$ be an arbitrary extension with L finite over K_0. If res_L is the homomorphism

$$H^2(\ |K) \xrightarrow{\ \mathrm{res}_L\ } H^2(\ |L)$$

(cf. §1, p. 66), then we have to show that

$$\mathrm{inv}_L \circ \mathrm{res}_L = [L : K] \cdot \mathrm{inv}_K$$

(cf. the remarks made at the end of (1.3)). If $c \in H^2(\ |K)$, we can assume by (5.2) that $c \in H^2(L'|K)$, where $L'|K$ is unramified. Then $N = L \cdot L'|L$ is also unramified, and $\mathrm{res}_L c \in H^2(N|L) \subseteq H^2(\ |L)$. From Lemma (5.3) we obtain

$$\mathrm{inv}_L(\mathrm{res}_L c) = [L : K] \cdot \mathrm{inv}_K c,$$

which proves our claim.

Theorem (5.6.) allows us to apply the theory of abstract class formations developed in §1; we consider the general results established there once more in their special form for the case at hand.

For every normal extension $L|K$ we have the **fundamental class**

$$u_{L|K} \in H^2(L|K) \text{ with the invariant } \mathrm{inv}_{L|K}\, u_{L|K} = \tfrac{1}{[L:K]} + \mathbb{Z}.$$

The **Main Theorem of Local Class Field Theory** is the statement

(5.7) Theorem. *Let* $L|K$ *be a normal extension. Then the homomorphism*

$$u_{L|K} \cup : H^q(G_{L|K}, \mathbb{Z}) \longrightarrow H^{q+2}(L|K)$$

is bijective for every integer q.

For $q = 1, 2$ it follows that (cf. (1.8))

(5.8) Corollary. $H^3(L|K) = 1$ *and* $H^4(L|K) = \chi(G_{L|K})$.

For $q = -2$ we get the **local reciprocity law**:

(5.9) Theorem. *For every normal extension $L|K$ we have the isomorphism*

$$G_{L|K}^{ab} \cong H^{-2}(G_{L|K}, \mathbb{Z}) \xrightarrow{u_{L|K} \cup} H^0(L|K) = K^\times / N_{L|K} L^\times.$$

The inverse isomorphism induces an exact sequence

$$1 \longrightarrow N_{L|K} L^\times \longrightarrow K^\times \xrightarrow{(\ ,L|K)} G_{L|K}^{ab} \longrightarrow 1,$$

where $(\ , L|K)$ is the **norm residue symbol**. When passing to subfields, extension fields and conjugate fields, the norm residue symbol behaves as follows:

(5.10) Proposition. *Let $N \supseteq L \supseteq K$ be extensions of K with $N|K$ normal. Then the following diagrams are commutative (cf. (1.11)):*

a)
$$
\begin{array}{ccc}
K^\times & \xrightarrow{(\ ,N|K)} & G_{N|K}^{ab} \\
{\scriptstyle \mathrm{id}}\downarrow & & \downarrow{\scriptstyle \pi} \\
K^\times & \xrightarrow{(\ ,L|K)} & G_{L|K}^{ab},
\end{array}
$$

b)
$$
\begin{array}{ccc}
K^\times & \xrightarrow{(\ ,N|K)} & G_{N|K}^{ab} \\
{\scriptstyle \mathrm{incl}}\downarrow & & \downarrow{\scriptstyle \mathrm{Ver}} \\
L^\times & \xrightarrow{(\ ,N|L)} & G_{N|L}^{ab},
\end{array}
$$

c)
$$
\begin{array}{ccc}
L^\times & \xrightarrow{(\ ,N|L)} & G_{N|L}^{ab} \\
{\scriptstyle N_{L|K}}\downarrow & & \downarrow{\scriptstyle \kappa} \\
K^\times & \xrightarrow{(\ ,N|K)} & G_{N|K}^{ab},
\end{array}
$$

d)
$$
\begin{array}{ccc}
K^\times & \xrightarrow{(\ ,N|K)} & G_{N|K}^{ab} \\
{\scriptstyle \sigma}\downarrow & & \downarrow{\scriptstyle \sigma^*} \\
\sigma K^\times & \xrightarrow{(\ ,\sigma N|\sigma K)} & G_{\sigma N|\sigma K}^{ab}.
\end{array}
$$

Here in diagram a) the extension $L|K$ is also assumed to be normal, and in diagram d) σ denotes an element in G_{K_0}.

By (1.10) the norm residue symbol and the invariant map are related:

(5.11) Lemma. *Let $L|K$ be a normal extension, $a \in K^\times$, and $\bar{a} = a \cdot N_{L|K} L^\times \in H^0(L|K)$. If $\chi \in \chi(G_{L|K}^{ab}) = \chi(G_{L|K}) = H^1(G_{L|K}, \mathbb{Q}/\mathbb{Z})$, then*

$$\chi(a, L|K) = \mathrm{inv}_{L|K}(\bar{a} \cup \delta\chi) \in \tfrac{1}{[L:K]}\mathbb{Z}/\mathbb{Z},$$

where $\delta\chi$ is the image of χ under $H^1(G_{L|K}, \mathbb{Q}/\mathbb{Z}) \xrightarrow{\delta} H^2(G_{L|K}, \mathbb{Z})$.

For unramified extensions $L|K$ we gave in (4.8)) an explicit description of the the norm residue symbol $(\ , L|K)$ in terms of the Frobenius automorphism:

$$(a, L|K) = \varphi_{L|K}^{v_K(a)}.$$

It is very important to have such an explicit representation in case of ramified extensions as well. Concerning this question, a fairly general result has been obtained in recent years by J. LUBIN and J. TATE. We will consider this representation in §7.

We extract the following consequence of (4.8):

(5.12) Theorem. *If $L|K$ is an abelian extension, then the norm residue symbol $(\ ,L|K)$ maps the unit group U_K onto the inertia group of $G_{L|K}$ and the principal unit group U_K^1 onto the ramification group.*

Proof. Let L_τ be the inertia field between K and L, $f = [L_\tau : K]$ and G_τ the inertia group of $G_{L|K}$, thus the fixed group of L_τ. If $u \in U_K$, then by (5.10 a) we have $\pi(u, L|K) = (u, L|K) \cdot G_\tau = (u, L_\tau|K) = \varphi_{L_\tau|K}^{v_K(u)} = 1$, and therefore $(u, L|K) \in G_\tau$. Conversely, let $\tau \in G_\tau$ and $a \in K^\times$ with $(a, L|K) = \tau$. Then

$$\pi(a, L|K) = (a, L|K) \cdot G_\tau = 1, \text{ therefore } (a, L_\tau|K) = \varphi_{L_\tau|K}^{v_K(a)} = 1,$$

i.e., $v_K(a) \equiv 0 \bmod f$. If we choose a $b \in L^\times$ with $v_L(b) = \frac{1}{f}v_K(a)$, then

$$v_L(N_{L|K}b) = e \cdot v_K(N_{L|K}b) = [L : K] \cdot v_L(b) = e \cdot v_K(a),$$

therefore $v_K(a) = v_K(N_{L|K}b)$, $a = u \cdot N_{L|K}b$ with $u \in U_K$. From this we have $(a, L|K) = (u, L|K) = \tau$, i.e., U_K is mapped onto the entire inertia group.

Observing that $(U_K^n, L|K) = 1$ for sufficiently large n, we conclude that the ramification group G_v, which is the only p-Sylow subgroup of G_τ, is the image of the p-Sylow subgroup U_K^1/U_K^n of U_K/U_K^n (cf. (3.2)).

It is possible to strengthen Theorem (5.12) by showing that the higher principal unit groups U_K^n, when suitably numbered, are mapped onto the higher ramification groups of $G_{L|K}$. For this we refer to [42], XV, §2, Cor. 3.

To end, we briefly discuss the universal norm residue symbol of our class formation (cf. §1, p. 76). For each abelian extension $L|K$ we have the map

$$K^\times \xrightarrow{(\ ,L|K)} G_{L|K}.$$

By forming the projective limit

$$G_K^{\mathrm{ab}} = \varprojlim G_{L|K} \qquad (L|K \text{ abelian})$$

we get, for every $a \in K^\times$, the element

$$(a, K) = \varprojlim(a, L|K) \in G_K^{\mathrm{ab}}$$

in the Galois group G_K^{ab} of the maximal abelian extension of K.

(5.13) Theorem. *The universal residue symbol defines an injective homomorphism*

$$K^\times \xrightarrow{(\ ,K)} G_K^{ab}.$$

Proof. By (1.15) the intersection $D_K = \bigcap_L N_{L|K} L^\times$ is the kernel of $(\ ,K)$. By (3.7) the groups of m-th powers $(K^\times)^m$ are of finite index, and therefore are norm groups by Theorem (6.3), which is proved in the next section. Therefore $D_K \subseteq \bigcap_{m=1}^\infty (K^\times)^m = 1$.

If $a \in K^\times$, then the restriction of $(a, K) \in G_K^{ab}$ to the inertia field $T|K$ gives the universal norm residue symbol of the class formation of the unramified extensions of K discussed in §4 (cf. §4, p. 87). By (4.10) we thus have

$$(a, K)\big|_T = (a, T|K) = \varphi_K^{v_K(a)} \in G_{T|K},$$

where $\varphi_K = \varprojlim_{L|K \text{ unram.}} \varphi_{L|K} \in G_{T|K}$ denotes the universal Frobenius automorphism.

In global class field theory we have to consider besides the $\mathfrak{p}$-adic number fields also the field $\mathbb{R}$ of real numbers. There is also a reciprocity law over the reals, which is so simple, however, that we can explain it with only a few words.

The field $\mathbb{R}$ has only one algebraic extension, namely the field $\mathbb{C}$ of complex numbers. The pair $(G_{\mathbb{C}|\mathbb{R}}, \mathbb{C}^\times)$ constitutes a class formation in a trivial way: The group

$$H^2(G_{\mathbb{C}|\mathbb{R}}, \mathbb{C}^\times) \cong H^0(G_{\mathbb{C}|\mathbb{R}}, \mathbb{C}^\times) = \mathbb{R}^\times / N_{\mathbb{C}|\mathbb{R}} \mathbb{C}^\times$$

is cyclic of order 2, since an element $a \in \mathbb{R}^\times$ is a norm from $\mathbb{C}$ if and only if $a > 0$. The invariant map

$$\mathrm{inv}_{\mathbb{C}|\mathbb{R}} : H^2(G_{\mathbb{C}|\mathbb{R}}, \mathbb{C}^\times) \longrightarrow \tfrac{1}{2}\mathbb{Z}/\mathbb{Z}$$

is defined in the obvious way, and the norm residue symbol $(\ , \mathbb{C}|\mathbb{R})$ is characterized by the equation

$$(a, \mathbb{C}|\mathbb{R})(\sqrt{-1}) = (\sqrt{-1})^{\mathrm{sgn}(a)},$$

since $(a, \mathbb{C}|\mathbb{R})$ is either the identity or the conjugation map, depending on whether a is a norm or not, i.e., whether $a > 0$ or $a < 0$.

§ 6. The Existence Theorem

From the abstract class field theory of §1 we see (cf. (1.14)) that the norm groups in a $\mathfrak{p}$-adic number field K correspond bijectively to the abelian extensions of K:

(6.1) Theorem. *Let K be a $\mathfrak{p}$-adic number field. Then the correspondence*

$$L \longmapsto I_L = N_{L|K} L^\times \subseteq K^\times$$

gives an inclusion reversing isomorphism between the lattice of abelian extensions $L|K$ and the lattice of norm groups in $K^\times$. Every group containing a norm group is again a norm group.

The last theorem shows that the structure of the abelian extensions of K is reflected in the multiplicative group $K^\times$, and naturally leads to the question of how the norm groups in $K^\times$ can be characterized by intrinsic properties of $K^\times$. This is done by the following so-called **Existence Theorem**:

(6.2) Theorem. *The norm groups of $K^\times$ are precisely the open (and thus closed) subgroups of finite index.*

Proof. By the reciprocity law (5.9) every norm group $I_L \subseteq K^\times$ has finite index in $K^\times$. If m is this index, then clearly $(K^\times)^m \subseteq I_L$. By (3.6) $(K^\times)^m$ is open, hence I_L is open as the union of (open) cosets of $(K^\times)^m$ in I_L.

Conversely, let $I \subseteq K^\times$ be an open subgroup of finite index m in $K^\times$. Then $(K^\times)^m \subseteq I$ and by (6.1) I is a norm group if $(K^\times)^m$ is a norm group. We show this first in case that K contains the m-th roots of unity. For each $a \in K^\times$ we form the field $L_a = K(\sqrt[m]{a})$, and set

$$L = \bigcup_{a \in K^\times} L_a.$$

Then $L|K$ is a finite abelian extension, because $K^\times / (K^\times)^m$ is finite (cf. (3.7)), and therefore there are only finitely many distinct fields among the L_a. We now claim that

$$(K^\times)^m = I_L = \bigcap_{a \in K^\times} I_{L_a} \qquad ^{12)}.$$

The degree $[L_a : K] = [K(\sqrt[m]{a}) : K] = d$ is obviously a divisor of m, hence the inclusion $(K^\times)^d \subseteq I_{L_a}$ implies $(K^\times)^m \subseteq I_{L_a}$ for all a. Therefore $(K^\times)^m \subseteq I_L$.

On the other hand, the theory of Kummer extensions (cf. Part III, §1, p. 115, (1.3)) gives an isomorphism between the factor group $K^\times / (K^\times)^m$ and the character group of the Galois group $G_{L|K}$, so that by (5.9)

$$(K^\times : (K^\times)^m) = |G_{L|K}| = (K^\times : I_L).$$

Thus $(K^\times)^m = I_L$, and therefore $(K^\times)^m$ is a norm group. If K does not contain the m-th roots of unity, let K_1 be the extension obtained by adjoining the m-th roots to K. From the above we know that $(K_1^\times)^m$ is the norm group of an extension $L|K_1 : (K_1^\times)^m = N_{L|K_1} L^\times$. Let $\tilde{L}$ be the smallest normal extension of K containing L. Then

---

$^{12)}$ The right equation follows from (6.1) because $L = \bigcup_{a \in K^\times} L_a$.

$$N_{\tilde{L}|K}\tilde{L}^\times = N_{K_1|K}(N_{\tilde{L}|K_1}\tilde{L}^\times) \subseteq N_{K_1|K}(N_{L|K_1}L^\times) = N_{K_1|K}((K_1^\times)^m)$$
$$= (N_{K_1|K}K_1^\times)^m \subseteq (K^\times)^m.$$

Hence $(K^\times)^m$ is a group containing the norm group $N_{\tilde{L}|K}\tilde{L}^\times$, thus by (6.1) $(K^\times)^m$ is a norm group itself, and our Existence Theorem is proved.

Theorem (6.2) is called the Existence Theorem because its crucial assertion is that given an open subgroup I of finite index in $K^\times$, there exists an abelian extension $L|K$ whose norm group $N_{L|K}L^\times = I$. This field L is uniquely determined and is called the **class field** associated with I.

It is clear that the open subgroups of finite index in $K^\times$ are also closed of finite index, and vice versa, since the complement of a subgroup of finite index in $K^\times$ consists of its finitely many cosets. More generally, we have

(6.3) Theorem. *If I is a subgroup of $K^\times$, then the following conditions are equivalent:*

(i)   *I is a norm group,*
(ii) *I is open of finite index,*
(iii) *I is is closed of finite index,*
(iv) *I has finite index.*

Proof.   The conditions (i), (ii), (iii) are equivalent by (6.2) and our remark above. Furthermore, (iv) is equivalent to (ii), since a subgroup I of finite index m contains the open group $(K^\times)^m$, and is therefore open.

Apart from the topological characterization of norm groups given by Theorem (6.3), we also have the following description of these groups, which is of an arithmetic nature (cf. also (4.9)).

(6.4) Theorem. *The norm groups of $K^\times$ are precisely the groups containing*
$$U_K^n \times (\pi^f), \qquad n = 0, 1, 2, \ldots, \quad f = 1, 2, \ldots$$
Here $U_K^0 = U_K$, π is a prime element of K, and (π^f) is the subgroup of $K^\times$ generated by π^f.

Proof.   Every group $U_K^n \times (\pi^f)$ has finite index in $K^\times = U_K^0 \times (\pi)$, and is therefore a norm group by (6.3), hence has only norm groups containing it.

Conversely, if I is a norm group, it is open. Since the U_K^n form a fundamental system of neighborhoods of $1 \in K^\times$, there is a U_K^n with $U_K^n \subseteq I$. If π is a prime element and f the index $(K^\times : I)$, then $\pi^f \in I$, and thus $U_K^n \times (\pi^f) \subseteq I$.

We will give a more detailed account of norm groups in the next section.

§ 7. Explicit Determination of the Norm Residue Symbol [13)

In Theorem (4.8) we have given an explicit description of the norm residue symbol for unramified extensions via the Frobenius automorphism. In this section we derive such an explicit formula for the norm residue symbol in certain special totally ramified extensions. Since these extensions together with the unramified extensions generate the maximal abelian field, this allows us to explicitly determine the universal norm residue symbol.

Let K be a $\mathfrak{p}$-adic number field, $\mathcal{o}$ the ring of integers of K, π a prime element and $q = (\mathcal{o} : \pi\mathcal{o})$ the number of elements in the residue field $\overline{K}$.

We consider the set ξ_π of all power series $f(Z) \in \mathcal{o}[[Z]]$ such that

$$f(Z) \equiv \pi \cdot Z \text{ mod degree } 2 \quad \text{and} \quad f(Z) \equiv Z^q \text{ mod } \pi.$$

Two power series are called congruent mod degree n (resp. mod π) if their terms of degree less than n coincide (resp. when their coefficients are congruent mod π). The simplest example of a power series in ξ_π is the polynomial $f(Z) = \pi \cdot Z + Z^q$, which one may regard as a standard model. Let

$$f^n(Z) = f(f(\cdots f(Z)\cdots)) \in \mathcal{o}[[Z]]$$

be the power series obtained by n-fold substitution in $f(Z)$, where $f^0(Z) = Z$.

Let $\Lambda_{f,n}$ be the set of those elements λ of positive valuation in the algebraic closure Ω of K with $f^n(\lambda) = 0$. We consider the fields

$$L_{f,n} = K(\Lambda_{f,n}), \quad n = 1, 2, \ldots$$

Because

$$f^n(Z) = f(f^{n-1}(Z)) = f^{n-1}(Z) \cdot \phi_n(Z), \quad \phi_n(Z) \in \mathcal{o}[[Z]],$$

it is immediately clear that $\Lambda_{f,n-1} \subseteq \Lambda_{f,n}$, and therefore that

$$L_{f,n-1} \subseteq L_{f,n}, \quad n = 1, 2, \ldots$$

We set $\Lambda_f = \bigcup_{n=1}^\infty \Lambda_{f,n}$ and $L_f = K(\Lambda_f) = \bigcup_{n=1}^\infty L_{f,n}$.

We will show that the extensions $L_{f,n}|K$ are abelian and totally ramified, and that they are associated with the norm groups $U_K^n \times (\pi)$ (cf. (6.4)). The essential idea here is to use certain power series to make the zero set $\Lambda_{f,n}$ an $\mathcal{o}$-module in such a way that multiplication of $\Lambda_{f,n}$ by a unit $u \in \mathcal{o}$ produces a permutation of $\Lambda_{f,n}$, which induces a K-automorphism of $L_{f,n}|K$, namely the automorphism $(u^{-1}, L_{f,n}|K)$.

[13)] In this section we follow [34]. For Part III only Theorem (7.16) is used.

(7.1) Lemma. *Let* $f(Z), g(Z) \in \xi_\pi$ *and* $L(X_1, \ldots, X_n) = \sum_{i=1}^n a_i X_i$ *be a linear form with coefficients* $a_i \in \mathcal{O}$. *Then there is a uniquely determined power series* $F(X_1, \ldots, X_n)$ *with coefficients in* $\mathcal{O}$ *with the properties*

$$F(X_1, \ldots, X_n) \equiv L(X_1, \ldots, X_n) \quad \text{mod degree } 2,$$
$$f(F(X_1, \ldots, X_n)) = F(g(X_1), \ldots, g(X_n)).$$

Proof. We set $X = (X_1, \ldots, X_n)$ and $g(X) = (g(X_1), \ldots, g(X_n))$. One immediately verifies that if $F(X)$ is a power series and $F_r(X) \in \mathcal{O}[X]$ are its truncations consisting of all terms of $F(X)$ of degree at most r, then $F(X)$ is a solution of the above problem if and only if $F(X) \equiv L(X)$ mod degree 2, thus $F_1(X) = L(X)$, and for every r the following congruence holds

$$(*) \qquad f(F_r(X)) \equiv F_r(g(X)) \quad \text{mod degree } (r+1).$$

For $r = 1$, i.e., for $F_1(X) = L(X)$ this is true. If we have found a unique $F_r(X) \in \mathcal{O}[X]$ satisfying condition $(*)$, and we set $F_{r+1}(X) = F_r(X) + \Delta_{r+1}(X)$ with a homogeneous form Δ_{r+1} of degree $r + 1$, it follows from

$$f(F_{r+1}(X)) \equiv f(F_r(X)) + \pi \cdot \Delta_{r+1}(X) \qquad \text{mod degree } (r+2),$$
$$F_{r+1}(g(X)) \equiv F_r(g(X)) + \pi^{r+1} \cdot \Delta_{r+1}(X) \qquad \text{mod degree } (r+2),$$

that for Δ_{r+1} the congruence

$$\Delta_{r+1}(X) \equiv \frac{f(F_r(X)) - F_r(g(X))}{\pi^{r+1} - \pi} \quad \text{mod degree } (r+2)$$

must be satisfied. Hence we obtain Δ_{r+1} is a unique way as the first truncation, i.e., as the homogeneous form of $(r + 1)$-th degree of the power series $\big(f(F_r(X)) - F_r(g(X))\big)/(\pi^{r+1} - \pi)$. Because

$$f(F_r(X)) - F_r(g(X)) \equiv (F_r(X))^q - F_r(X^q) \equiv 0 \quad \text{mod } \pi$$

the form Δ_{r+1} has integral coefficients, and hence so has $F_{r+1} = F_r + \Delta_{r+1}$. This shows existence and uniqueness of the series $F(X) = \lim_{r \to \infty} F_r(X)$.

Remark. The proof actually shows that F is the only power series in every field containing $\mathcal{O}$ which satisfies the equations of the lemma.

For us, the cases $L(X, Y) = X + Y$ and $L(Z) = aZ$, $a \in \mathcal{O}$, are important. If $f \in \xi_\pi$, let $F_f(X, Y)$ be the uniquely determined solution of the equations

$$F_f(X, Y) = X + Y \quad \text{mod degree } 2,$$
$$f(F_f(X, Y)) = F_f(f(X), f(Y)).$$

Moreover, for every $a \in \mathcal{O}$ and $f, g \in \xi_\pi$ let the series $a_{f,g}(Z) \in \mathcal{O}[[Z]]$ be the uniquely determined solution of

$$a_{f,g}(Z) \equiv aZ \quad \text{mod degree } 2,$$
$$f(a_{f,g}(Z)) = a_{f,g}(g(Z)).$$

For simplicity, we write a_f for $a_{f,f}$. The following proposition shows that $F_f(X,Y)$ in a certain sense plays the role of "addition," while a_f corresponds to "multiplication."

(7.2) Proposition. Let $f, g, h \in \xi_\pi$ and $a, b \in \mathcal{O}$. Then

(1) $\qquad\qquad F_f(X,Y) = F_f(Y,X),$

(2) $\qquad F_f(F_f(X,Y),Z) = F_f(X,F_f(Y,Z)),$

(3) $\qquad a_{f,g}(F_g(X,Y)) = F_f(a_{f,g}(X),a_{f,g}(Y)),$

(4) $\qquad a_{f,g}(b_{g,h}(Z)) = (a \cdot b)_{f,h}(Z),$

(5) $\qquad (a+b)_{f,g}(Z) = F_f(a_{f,g}(Z),b_{f,g}(Z)),$

(6) $\qquad\qquad (\pi^n)_f(Z) = f^n(Z) , \; n = 0,1,2,\dots$

To prove these formulas, one shows that the left and right side of each equation are both solution of a problem as in (7.1), hence by uniqueness of the solution they are equal. We leave the details to the reader.

For $f = g = h$ we obtain from (1)–(6) the formal rules of an $\mathcal{O}$-module. Therefore we call F_f a **formal Lie $\mathcal{O}$-module**. From such a formal Lie $\mathcal{O}$-module we obtain an ordinary $\mathcal{O}$-module by letting the variables X, Y, Z attain values from a domain in which the power series converge. If L is an arbitrary algebraic extension of K, then the prime ideal $\mathfrak{p}_L$ of the elements of positive valuation in L represents such a domain. In fact, if $x_1, \dots, x_n \in \mathfrak{p}_L$ and $G(X_1, \dots, X_n) \in \mathcal{O}[[X_1, \dots, X_n]]$, then the series $G(x_1, \dots, x_n)$ converges and gives an element in $\mathfrak{p}_L$ if the constant term of G is zero[14]. We have so

(7.3) Proposition. Let $f \in \xi_\pi$, and let L be an algebraic extension of K. Then the set $\mathfrak{p}_L$ is an $\mathcal{O}$-module with addition and multiplication defined by

$$x + y = F_f(x,y) \quad \text{and} \quad a \cdot x = a_f(x), \; x, y \in \mathfrak{p}_L, \; a \in \mathcal{O};$$

we write $\mathfrak{p}_L^{(f)}$ for this $\mathcal{O}$-module.

Obviously, the additive inverse of x is $(-1)_f(x)$. One needs to be careful not confuse the operations $\mathfrak{p}_L^{(f)}$ with the ordinary operations on the $\mathcal{O}$-module.

(7.4) Proposition. The set of zeros $\Lambda_{f,n}$ of $f^n(x)$ is a submodule of $\mathfrak{p}_{L_{f,n}}^{(f)}$.

Proof. The set of zeros $\Lambda_{f,n}$ is the annihilator of the element $\pi^n \in \mathcal{O}$, since

$$\Lambda_{f,n} = \{\lambda \in \mathfrak{p}_{L_{f,n}} \mid f^n(\lambda) = (\pi^n)_f(\lambda) = 0\} = \{\lambda \in \mathfrak{p}_{L_{f,n}}^{(f)} \mid \pi^n \cdot \lambda = 0\}.$$

---

[14] $G(x_1, \dots, x_n)$ converges in the finite (and thus complete) extension of K generated by $x_1, \dots, x_n$.

(7.5) Proposition. *Let* $f, g \in \xi_\pi$ *and* $a \in \mathcal{o}$. *Then the map*

$$\lambda \longmapsto a_{g,f}(\lambda)$$

yields a homomorphism from $\Lambda_{f,n}$ *to* $\Lambda_{g,n}$. *This homomorphism is an isomorphism if* a *is a unit in* $\mathcal{o}$.

Proof. This follows immediately from the formulas in (7.2). By (3) and (4) this map is a homomorphism. If a is a unit, then (4) and (6) imply that

$$(a^{-1})_{f,g}(a_{g,f}(\lambda)) = 1_f(\lambda) = \lambda \text{ for } \lambda \in \Lambda_{f,n},$$

and in reverse order

$$a_{g,f}((a^{-1})_{f,g}(\lambda)) = 1_g(\lambda) = \lambda \text{ for } \lambda \in \Lambda_{g,n}.$$

Hence the map $a_{g,f} : \Lambda_{f,n} \to \Lambda_{g,n}$ is a bijection with inverse map $(a^{-1})_{f,g}$. $\square$

(7.6) Corollary. *Let* $f \in \xi_\pi$. *Then we have an isomorphism of* $\mathcal{o}$-*modules*

$$\Lambda_{f,n} \cong \mathcal{o}/\pi^n \cdot \mathcal{o}.$$

Proof. If $f, g \in \xi_\pi$, then the map $1_{g,f} : \Lambda_{f,n} \to \Lambda_{g,n}$ is an isomorphism by (7.5). Thus it suffices to consider the module $\Lambda_{f,n}$ with $f(Z) = \pi Z + Z^q \in \xi_\pi$.

The $\mathcal{o}$-module $\Lambda_{f,1}$ consists of the zeros of the equation $f(Z) = \pi Z + Z^q = 0$, thus has q elements, and is therefore a one-dimensional vector space over the field $\mathcal{o}/\pi \cdot \mathcal{o}$. For $n = 1$ the result follows from the isomorphism $\Lambda_{f,1} \cong \mathcal{o}/\pi \mathcal{o}$.

Assume that $\Lambda_{f,n} \cong \mathcal{o}/\pi^n \cdot \mathcal{o}$. By (7.5) the element π defines the homomorphism $\pi_f : \Lambda_{f,n+1} \to \Lambda_{f,n}$, from which we obtain the exact sequence

$$0 \longrightarrow \Lambda_{f,1} \longrightarrow \Lambda_{f,n+1} \xrightarrow{\pi_f} \Lambda_{f,n} \longrightarrow 0.$$

To see this, note first that $\pi_f(\lambda) \in \Lambda_{f,n}$ for $\lambda \in \Lambda_{f,n+1}$, because $f^n(\pi_f(\lambda)) = f^n(f(\lambda)) = f^{n+1}(\lambda) = 0$. If $\lambda \in \Lambda_{f,n}$ and $\lambda^*(\in \Omega)$ is a root of the equation $f(Z) - \lambda = Z^q + \pi Z - \lambda = 0$, then $\lambda^* \in \Lambda_{f,n+1}$ because $f^{n+1}(\lambda^*) = f^n(f(\lambda^*)) = f^n(\lambda) = 0$, thus $\pi_f(\lambda^*) = f(\lambda^*) = \lambda$, and π_f is surjective. The kernel of π_f consists of the elements λ with $\pi_f(\lambda) = f(\lambda) = 0$, i.e., the elements of the module $\Lambda_{f,1} \subseteq \Lambda_{f,n+1}$.

Since $\Lambda_{f,1} \cong \mathcal{o}/\pi \cdot \mathcal{o}$ and $\Lambda_{f,n} \cong \mathcal{o}/\pi^n \cdot \mathcal{o}$, the order of $\Lambda_{f,n+1}$ is equal to q^{n+1}. If $\lambda \in \Lambda_{f,n+1}$ but $\lambda \notin \Lambda_{f,n}$, then $\pi^{n+1} \cdot \mathcal{o}$ is clearly the annihilator of λ. Hence the map $a \mapsto a \cdot \lambda$ gives an isomorphism between $\mathcal{o}/\pi^{n+1} \cdot \mathcal{o}$ and the $\mathcal{o}$-submodule of $\Lambda_{f,n+1}$ generated by λ, which must coincide with $\Lambda_{f,n+1}$, since $\mathcal{o}/\pi^{n+1} \cdot \mathcal{o}$ and $\Lambda_{f,n+1}$ both have order q^{n+1}. Therefore $\Lambda_{f,n+1} \cong \mathcal{o}/\pi^{n+1} \cdot \mathcal{o}$. $\square$

(7.7) Corollary. *Every automorphism of the* $\mathcal{o}$-*module* $\Lambda_{f,n}$ *is of the form* $u_f : \Lambda_{f,n} \to \Lambda_{f,n}$ *with a unit* $u \in U_K$. *The map* u_f *is the identity on* $\Lambda_{f,n}$ *if and only if* $u \in U_K^n$. *Thus the group* U_K/U_K^n *represents the full automorphism group of* $\Lambda_{f,n}$.

The proof follows easily from the isomorphism $\Lambda_{f,n} \cong o/\pi^n \cdot o$ and is left to the reader.

(7.8) Proposition. *The field $L_{f,n}$ depends only on the prime element π, but not on the choice of the power series $f \in \xi_\pi$.*

Proof. If $f, g \in \xi_\pi$ and $\lambda \in \Lambda_{f,n}$, then along with λ we have $1_{g,f}(\lambda) \in L_{f,n}$. Since $1_{g,f} : \Lambda_{f,n} \to \Lambda_{g,n}$ is surjective, $\Lambda_{g,n} \subseteq L_{f,n}$, i.e., $L_{g,n} \subseteq L_{f,n}$. By symmetry, $L_{f,n} \subseteq L_{g,n}$, and therefore $L_{f,n} = L_{g,n}$.

Because of this result we write $L_{\pi,n}$ for the field $L_{f,n}$, and set $L_\pi = \bigcup_{n=1}^{\infty} L_{\pi,n}$. We can always think of $L_{\pi,n}$ as generated by the roots of the polynomials $f^n(Z)$ with $f(Z) = \pi Z + Z^q \in \xi_\pi$, hence $L_{\pi,n} | K$ is a normal extension. We denote its Galois group by $G_{\pi,n}$. The projective limit $G_\pi = \varprojlim G_{\pi,n}$ is the Galois group of the extension $L_\pi | K$. Each element $\sigma \in G_{\pi,n}$ yields an automorphism of the o-module $\Lambda_{f,n}$ by the usual operation of $G_{\pi,n}$ on the set $\Lambda_{f,n} \subseteq L_{\pi,n}$. This is due to the fact that σ acts continuously on $L_{\pi,n}$, and that the operations of the o-module $\Lambda_{f,n}$ are defined by convergent power series whose coefficients lie in the base field K, and therefore are fixed by σ. On the other hand, by (7.7) every class $u \cdot U_K^n \in U_K/U_K^n$ yields the automorphism $u_f : \Lambda_{f,n} \to \Lambda_{f,n}$. We show

(7.9) Theorem. *For every $\sigma \in G_{\pi,n}$ there is a uniquely determined class $u \cdot U_K^n \in U_K/U_K^n$ such that $\sigma(\lambda) = u_f(\lambda)$, $\lambda \in \Lambda_{f,n}$. The map $\sigma \mapsto u \cdot U_K^n$ yields an isomorphism*

$$G_{\pi,n} \cong U_K/U_K^n.$$

Proof. Each $\sigma \in G_{\pi,n}$ induces an automorphism of the o-module $\Lambda_{f,n}$. Since by (7.7) U_K/U_K^n represents the full automorphism group of $\Lambda_{f,n}$, there is a (obviously unique) class $u \cdot U_K^n \in U_K/U_K^n$ with $\sigma(\lambda) = u_f(\lambda)$ for all $\lambda \in \Lambda_{f,n}$.

The map $\sigma \mapsto u \cdot U_K^n$ is injective, since the set $\Lambda_{f,n}$ generates the field $L_{\pi,n}$; thus it follows immediately from $\sigma(\lambda) = u_f(\lambda) = \lambda$ for all $\lambda \in \Lambda_{f,n}$ that $\sigma = 1$.

To prove $\sigma \mapsto u \cdot U_K^n$ is surjective, we show that the order of $G_{\pi,n}$ is not less than the order $q^{n-1}(q-1)$ of U_K/U_K^n (cf. (3.2)). We have

$$f^n(Z) = f(f^{n-1}(Z)) = f^{n-1}(Z) \cdot \phi_n(Z) \quad \text{with}$$
$$\phi_n(Z) = (f^{n-1}(Z))^{q-1} + \pi \in o[Z].$$

Since all coefficients of the polynomial $f^{n-1}(Z) = Z^{q^{n-1}} + \cdots + \pi^{n-1}Z$ have positive valuation, $\phi_n(Z)$ is an Eisenstein polynomial and as such irreducible over K. If λ is a root of $\phi_n(Z)$, and therefore a root of $f^n(Z)$, then $K(\lambda)$ is a totally ramified extension of $L_{\pi,n}$. Its degree $[K(\lambda) : K]$ is equal to the degree $q^n - q^{n-1} = q^{n-1}(q-1)$ of the polynomial $\phi_n(Z) = f^n(Z)/f^{n-1}(Z)$, thus equal to the order of $G_{\pi,n}$, which is at least $q^{n-1}(q-1) = |U_K/U_K^n|$. Therefore $G_{\pi,n} \cong U_K/U_K^n$ and $L_{\pi,n} = K(\lambda)$, where λ is a root of the Eisenstein polynomial $\phi_n(Z) = (f^{n-1}(Z))^{q-1} + \pi$. This completes the proof.

The proof of (7.9) also shows

(7.10) Theorem. *The extension $L_{\pi,n}|K$ is abelian and totally ramified of degree $q^{n-1}(q-1)$, and is generated by a root of the Eisenstein equation*

$$\phi_n(Z) = (f^{n-1}(Z))^{q-1} + \pi = 0.$$

The last statement shows that the prime element π is a norm for every extension $L_{\pi,n}|K$. In fact, if λ is a root of $\phi_n(Z)$, then $L_{\pi,n} = K(\lambda)$ and $\pi = N_{L_{\pi,n}|K}(-\lambda)$.

So far we have always fixed an arbitrarily chosen prime element π of K. Now we have to investigate what happens when we pass from π to another prime element π'. For this we use a lemma about the completion $\widehat{T}$ of the maximal unramified extension T over K. We again denote by φ the (universal) Frobenius automorphism of $T|K$, whose restriction to an unramified finite extension $L|K$ yields the Frobenius automorphism $\varphi_{L|K}$ (cf. §4, p. 88). If we think of φ as continuously extended to the completion $\widehat{T}$, then we have the

Sublemma. $U_{\widehat{T}}^{\varphi-1} = U_{\widehat{T}}$ and $(\varphi-1)\mathfrak{o}_{\widehat{T}} = \mathfrak{o}_{\widehat{T}}$.

Proof. It follows immediately from the definition of the Frobenius automorphism φ that the automorphism $\bar{\varphi}$ of the algebraic closure of the residue field $\overline{\widehat{T}} = \overline{T}$ induced by φ satisfies the following equations

$$(*) \qquad (\overline{\widehat{T}}^{\times})^{\bar{\varphi}-1} = \overline{\widehat{T}}^{\times} \quad \text{and} \quad (\bar{\varphi}-1)(\overline{\widehat{T}}^{+}) = \overline{\widehat{T}}^{+},$$

where $\overline{\widehat{T}}^{\times}$ and $\overline{\widehat{T}}^{+}$ are the multiplicative and additive group of $\overline{\widehat{T}}$. Furthermore

$$(**) \quad U_{\widehat{T}}/U_{\widehat{T}}^1 \cong \overline{\widehat{T}}^{\times}, \ U_{\widehat{T}}^n/U_{\widehat{T}}^{n+1} \cong \overline{\widehat{T}}^{+} \quad \text{and} \quad \mathfrak{o}_{\widehat{T}}/\mathfrak{p}_{\widehat{T}} \cong \mathfrak{p}_{\widehat{T}}^n/\mathfrak{p}_{\widehat{T}}^{n+1} \cong \overline{\widehat{T}}^{+}.$$

Now if $x \in U_{\widehat{T}}$, resp. $x \in \mathfrak{o}_{\widehat{T}}$, then $\bar{x} = \bar{\varphi}\bar{y}_1/\bar{y}_1 \in \overline{\widehat{T}}^{\times}$, resp. $\bar{x} = \bar{\varphi}\bar{y}_1 - \bar{y}_1 \in \overline{\widehat{T}}^{+}$, so that

$$x = \frac{\varphi y_1}{y_1}a_1, \ y_1 \in U_{\widehat{T}}, \ a_1 \in U_{\widehat{T}}^1, \quad \text{resp.} \quad x = \varphi y_1 - y_1 + a_1, \ y_1 \in \mathfrak{o}_{\widehat{T}}, \ a_1 \in \mathfrak{p}_{\widehat{T}}.$$

Because of $(*)$ and $(**)$ we obtain

$$a_1 = \frac{\varphi y_2}{y_2}a_2, \ y_2 \in U_{\widehat{T}}^1, \ a_2 \in U_{\widehat{T}}^2, \quad \text{resp.} \quad a_1 = \varphi y_2 - y_2 + a_2, \ y_2 \in \mathfrak{p}_{\widehat{T}}, \ a_2 \in \mathfrak{p}_{\widehat{T}}^2,$$

and therefore

$$x = \frac{\varphi(y_1 \cdot y_2)}{y_1 \cdot y_2} \cdot a_2, \quad \text{resp.} \quad x = \varphi(y_1 + y_2) - (y_1 + y_2) + a_2.$$

Continuing in this way, we have

$$x = \frac{\varphi(y_1 \cdots y_n)}{y_1 \cdots y_n} \cdot a_n, \ y_n \in U_{\widehat{T}}^{n-1}, \ a_n \in U_{\widehat{T}}^{n}, \quad \text{resp.}$$

$$x = \varphi(y_1 + \cdots + y_n) - (y_1 + \cdots + y_n) + a_n, \ y_n \in \mathfrak{p}_{\widehat{T}}^{n-1}, \ a_n \in \mathfrak{p}_{\widehat{T}}^{n},$$

and passing to the limit

$$x = \frac{\varphi y}{y}, \ y = \prod_{n=1}^{\infty} y_n \in U_{\widehat{T}}, \quad \text{resp.} \quad x = \varphi y - y, \ y = \sum_{n=1}^{\infty} y_n \in \mathcal{O}_{\widehat{T}}.$$

Using this sublemma, we now prove a result similar to (7.1).

(7.11) Lemma. Let π and $\pi' = u \cdot \pi$ $(u \in U_K)$ be two prime elements of K, and let $f \in \xi_\pi$, $f' \in \xi_{\pi'}$. Then there is a power series

$$\theta(Z) \equiv \varepsilon Z \mod \text{degree } 2, \quad \varepsilon \text{ a unit,}$$

with coefficients in the ring $\mathcal{O}_{\widehat{T}}$ of integers of $\widehat{T}$ with the following properties:

(1) $\qquad \theta^\varphi(Z) = \theta(u_f(Z)),$ [15)]

(2) $\qquad \theta(F_f(X,Y)) = F_{f'}(\theta(X), \theta(Y)),$

(3) $\qquad \theta(a_f(Z)) = a_{f'}(\theta(Z)) \quad \text{for all } a \in \mathcal{O}.$

Proof. By the sublemma we have $u = \varphi\varepsilon/\varepsilon$, $\varepsilon \in U_{\widehat{T}}$, and we set $\theta_1(Z) = \varepsilon Z$. We assume that we have constructed a polynomial $\theta_r(Z)$ of degree r such that

$$\theta_r^\varphi(Z) \equiv \theta_r(u_f(Z)) \mod \text{degree } (r+1),$$

and look for a polynomial $\theta_{r+1}(Z) = \theta_r(Z) + bZ^{r+1}$, which satisfies the same congruence with $r + 1$ instead of r. If we set $b = a \cdot \varepsilon^{r+1}$, we obtain for a the condition $a - \varphi a = c/(\varphi\varepsilon)^{r+1}$, where c is the coefficient of Z^{r+1} in the series $\theta_r^\varphi(Z) - \theta_r(u_f(Z))$. Because of the sublemma there always exists such an a; thus we obtain θ_{r+1} and therefore the series $\theta(Z) = \lim_{r \to \infty} \theta_r(Z)$ satisfying the condition $\theta^\varphi(Z) = \theta(u_f(Z))$.

[15)] θ^φ is the power series obtained from θ by applying φ to the coefficients of θ.

In order to obtain (2) and (3), we have to modify the θ we just constructed. Consider the series

$$h = \theta^\varphi \circ f \circ \theta^{-1} = \theta \circ u_f \circ f \circ \theta^{-1} = \theta \circ \pi'_f \circ \theta^{-1},$$

where the symbol $\circ$ stands for evaluation. Its coefficients lie in $\mathcal{O}_{\widehat{T}}$, and because $h^\varphi = \theta^\varphi \circ \pi'^\varphi_f \circ \theta^{-\varphi} = \theta^\varphi \circ f \circ u_f \circ \theta^{-\varphi} = h$, they even lie in $\mathcal{O}$, since an element in $\widehat{T}$ which is fixed by φ lies in K, as one can easily verify. From this we get

$$h(Z) \equiv \varepsilon \cdot \pi' \cdot \varepsilon^{-1} Z = \pi' Z \quad \text{mod degree 2} \quad \text{and}$$
$$h(Z) = \theta^\varphi(f(\theta^{-1}(Z))) \equiv \theta^\varphi(\theta^{-1}(Z)^q) \equiv \theta^\varphi(\theta^{-\varphi}(Z^q)) \equiv Z^q \quad \text{mod } \pi',$$

so that $h \in \xi_{\pi'}$. Now we replace θ by $1_{f',h} \circ \theta$; then (1) still holds for this modified θ, and we have $f' = \theta^\varphi \circ f \circ \theta^{-1} = \theta \circ \pi'_f \circ \theta^{-1}$.

For the proof of (2), we show that the series

$$F(X,Y) = \theta(F_f(\theta^{-1}(X), \theta^{-1}(Y)))$$

satisfies the conditions of (7.1), which characterize the series $F_{f'}(X,Y)$. It is clear that $F(X,Y) \equiv X + Y$ mod degree 2, a trivial calculation using the formula $f' = \theta \circ \pi'_f \circ \theta^{-1}$ shows that $F(f'(X), f'(Y)) = f'(F(X,Y))$, and by the remark at the end of (7.1) the coefficients of $F(X,Y)$ lie in $\mathcal{O}$.

Now (3) follows similarly, one shows that the series $\theta \circ a_f \circ \theta^{-1}$ satisfies the conditions of (7.1), which characterize the series $a_{f'}$.

(7.12) Corollary. Let π and $\pi' = u \cdot \pi$ be two prime elements in K, and let $f \in \xi_\pi$, $f' \in \xi_{\pi'}$. Then $\lambda \mapsto \theta(\lambda)$ yields an isomorphism of $\mathcal{O}$-modules

$$\Lambda_{f,n} \cong \Lambda_{f',n}.$$

Proof. Note first that if $\lambda \in \Lambda_{f,n}$, then $\theta(\lambda) \in \Lambda_{f',n}$, because

$$f'^n(\theta(\lambda)) = (\pi'^n)_{f'}(\theta(\lambda)) = \theta((u^n \cdot \pi^n)_f(\lambda)) = \theta(0) = 0.$$

That the map $\lambda \mapsto \theta(\lambda)$ is a homomorphism follows immediately from the formulas (2) and (3) of (7.11). If $\theta(\lambda) = 0$, then we necessarily have $\lambda = 0$, since otherwise $0 = \varepsilon + a_1\lambda + \cdots$, which is not possible because ε is a unit. Therefore the map is injective. It is also surjective, since by (7.6) both $\Lambda_{f,n}$ and $\Lambda_{f',n}$ are isomorphic to $\mathcal{O}/\pi^n \cdot \mathcal{O} = \mathcal{O}/\pi'^n \cdot \mathcal{O}$, and therefore have the same order.

For distinct prime elements π and π' of K, the fields $L_{\pi,n}$ and $L_{\pi',n}$ may very well be distinct. However, the previous corollary implies the following:

(7.13) Proposition. $T \cdot L_{\pi,n} = T \cdot L_{\pi',n}$.

Proof. By (7.12) $\Lambda_{f',n} = \theta(\Lambda_{f,n}) \subseteq \widehat{T \cdot L_{\pi,n}}$ (completion of $T \cdot L_{\pi,n}$). Since $\Lambda_{f',n}$ generates the field $L_{\pi',n}$, we have $\widehat{T \cdot L_{\pi',n}} \subseteq \widehat{T \cdot L_{\pi,n}}$, and thus by symmetry $\widehat{T \cdot L_{\pi',n}} = \widehat{T \cdot L_{\pi,n}}$. Therefore $T \cdot L_{\pi',n} = T \cdot L_{\pi,n}$, since both fields represent the algebraic closure of K in $\widehat{T \cdot L_{\pi,n}}$.

Since T is unramified and $L_{\pi,n}$ is totally ramified over K, $T \cap L_{\pi,n} = K$. Therefore the Galois group $G_{T \cdot L_{\pi,n}|K}$ of $T \cdot L_{\pi,n}|K$ is the direct product

$$G_{T \cdot L_{\pi,n}|K} = G_{T|K} \times G_{\pi,n}.$$

We now define a homomorphism

$$\omega_\pi : K^\times \longrightarrow G_{T \cdot L_{\pi,n}|K}$$

as follows: If $a = u \cdot \pi^m \in K^\times$, $u \in U_K$, then let

$$\omega_\pi(a)\big|_T = \varphi^m \in G_{T|K},$$
$$\omega_\pi(a)\big|_{L_{\pi,n}} = \sigma_u \in G_{\pi,n},$$

where σ_u is the automorphism on $L_{\pi,n}$ that corresponds by (7.9) to the class $u^{-1} \cdot U_K^n \in U_K/U_K^n$; in other words, the restriction $\omega_\pi(a)\big|_{L_{\pi,n}}$ is determined by $\omega_\pi(a)\lambda = (u^{-1})_f(\lambda)$, $\lambda \in \Lambda_{f,n}$.

We now come to the main goal of this section by showing that the homomorphism ω_π coincides with the one induced by the universal norm residue symbol $(\ ,K)$:

(7.14) Proposition. *For every $a \in K^\times$, we have*

$$\omega_\pi(a) = (a, K)|_{T \cdot L_{\pi,n}}.$$

Proof. Since the prime elements of K obviously generate $K^\times$, it suffices to prove the theorem for prime elements a. First let $a = \pi$. Then

$$\omega_\pi(\pi)|_T = \varphi = (\pi, T|K) = (\pi, K)|_T \qquad \text{(cf. (4.10))}$$

and, since by (7.10) π is a norm from $L_{\pi,n}$, we have

$$\omega_\pi(\pi)|_{L_{\pi,n}} = \sigma_1 = \mathrm{id}_{L_{\pi,n}} = (\pi, L_{\pi,n}|K) = (\pi, K)|_{L_{\pi,n}}.$$

Therefore

$$\omega_\pi(\pi) = (\pi, K)|_{T \cdot L_{\pi,n}}.$$

If $\pi' = u \cdot \pi$, $u \in U_K$, is another prime element of K, then $T \cdot L_{\pi,n} = T \cdot L_{\pi',n}$ by (7.13) and, we have again (cf. (4.10))

$$\omega_\pi(\pi')|_T = \varphi = (\pi', T|K) = (\pi', K)|_T.$$

Thus it remains to show that

$$\omega_\pi(\pi')|_{L_{\pi',n}} = (\pi', K)|_{L_{\pi',n}}.$$

Now $(\pi', K)|_{L_{\pi',n}} = (\pi', L_{\pi',n}|K) = \mathrm{id}_{L_{\pi',n}}$, since by (7.10) π' is a norm of $L_{\pi',n}$. This means that we have to verify that

$$\omega_\pi(\pi')|_{L_{\pi',n}} = \mathrm{id}_{L_{\pi',n}},$$

in other words, we have to show that $\omega_\pi(\pi')\lambda' = \lambda'$ for $\lambda' \in \Lambda_{f',n}$, where $f' \in \xi_{\pi'}$. By (7.12) $\Lambda_{f',n} = \theta(\Lambda_{f,n})$, and our claim will follow if we can show

$$\omega_\pi(\pi')\theta(\lambda) = \theta(\lambda) \quad \text{for } \lambda \in \Lambda_{f,n}.$$

For this, we consider

$$\omega_\pi(\pi') = \omega_\pi(u \cdot \pi) = \omega_\pi(u) \circ \omega_\pi(\pi).$$

From the above we have the following identities

$$\omega_\pi(\pi)\lambda = \lambda \ (\lambda \in \Lambda_{f,n}), \quad \omega_\pi(\pi)|_T = \varphi, \quad \omega_\pi(u)|_T = \mathrm{id}_T.$$

If we think of the last two automorphisms as continuously extended to $\widehat{T}$, then we obtain from (7.11) the formula

$$\omega_\pi(\pi')\theta(\lambda) = (\omega_\pi(u) \circ \omega_\pi(\pi))\theta(\lambda) = \omega_\pi(u)\theta^\varphi(\lambda) = \theta^\varphi(\omega_\pi(u)\lambda)$$
$$= \theta^\varphi((u^{-1})_f(\lambda)) = \theta(\lambda),$$

which completes the proof.

We can describe the norm residue symbol $(\ ,L_{\pi,n}|K)$ of the abelian and totally ramified extension $L_{\pi,n}|K$ as follows:

(7.15) Theorem. If $a = u \cdot \pi^m \in K^\times$, $u \in U_K$, then

$$(a, L_{\pi,n}|K)\lambda = (u^{-1})_f(\lambda) \quad \text{for all } \lambda \in \Lambda_{f,n} \subseteq L_{\pi,n}.$$

The norm group of the extension $L_{\pi,n}|K$ is the group $U_K^n \times (\pi)$.

Proof. It follows from (7.14) that $(a, K)|_{T \cdot L_{\pi,n}} = \omega_\pi(a)$. Therefore $(a, K)\lambda = (a, L_{\pi,n}|K)\lambda = \omega_\pi(a)\lambda = (u^{-1})_f(\lambda)$. Thus an element $a = u \cdot \pi^m \in K^\times$, $u \in U_K$, is a norm of the extension $L_{\pi,n}|K$ if and only if $(a, L_{\pi,n}|K)\lambda = (u^{-1})_f(\lambda) = \lambda$ for all $\lambda \in \Lambda_{f,n}$. By (7.7) this is equivalent to $u \in U_K^n$, i.e., to $a \in U_K^n \times (\pi)$.

As an application of (7.15) we discuss an example which can be considered the starting point of the material presented in this section; this example is also important for global class field theory (cf. the proof of (5.5) in Part III).

Let $K = \mathbb{Q}_p$ be the field of p-adic numbers. Then p is a prime element in K, and we choose for $f \in \xi_p$ the polynomial

$$f(Z) = (1 + Z)^p - 1 = pZ + \binom{p}{2}Z^2 + \cdots + Z^p,$$

so that

$$f^n(Z) = (1 + Z)^{p^n} - 1.$$

The zero set $\Lambda_{f,n}$ consists of the elements $\lambda = \zeta - 1$, where ζ runs through the p^n-th roots of unity. Thus the field $L_{p,n}$ is precisely the field of p^n-th roots of unity over $\mathbb{Q}_p$. Hence we have:

(7.16) Theorem. *Let* $a = u \cdot p^m \in \mathbb{Q}_p^\times$, *$u$ a unit, and let ζ be a primitive p^n-th root of unity. Then*

$$(a, \mathbb{Q}_p(\zeta)|\mathbb{Q}_p)\zeta = \zeta^r,$$

where r is a positive integer which is mod p^n determined by the congruence

$$r \equiv u^{-1} \bmod p^n.$$

Proof. Set $\lambda = \zeta - 1 \in \Lambda_{f,n}$. Then $r \cdot u \equiv 1 \bmod p^n$, and by (7.15) and (7.7)

$$(a, \mathbb{Q}_p(\zeta)|\mathbb{Q}_p)\lambda = (u^{-1})_f(\lambda) = r_f(\lambda).$$

On the other hand,
$$r_f(Z) = (1+Z)^r - 1,$$

since this polynomial obviously satisfies the conditions of the definition

$$r_f(Z) \equiv r\, Z \bmod \text{degree} \, 2 \quad \text{and} \quad f(r_f(Z)) = r_f(f(Z)).$$

Therefore

$$(a, \mathbb{Q}_p(\zeta)|\mathbb{Q}_p)\zeta = r_f(\lambda) + 1 = r_f(\zeta - 1) + 1 = \zeta^r.$$

After this example we return to the general case. In (4.9) we have shown that the norm groups of the unramified extensions $L|K$ are the groups $U_K \times (\pi^f)$. We now characterize the norm groups in the totally ramified case.

(7.17) Theorem. *The norm groups of the totally ramified (abelian) extensions $L|K$ are precisely the groups which contain the groups*

$$U_K^n \times (\pi) \qquad (\pi \text{ prime element}).$$

Proof. By (7.15), a group containing $U_K^n \times (\pi)$ belongs to a subfield of $L_{\pi,n}$, and thus to a totally ramified extension $L|K$. On the other hand, a totally ramified extension $L|K$ will be generated by a root λ of an Eisenstein equation

$$X^e + \cdots + \pi = 0,$$

where the prime element π is the norm of the element $\pm\lambda$. Therefore we have $(\pi) \subseteq N_{L|K}L^\times$. Since $N_{L|K}L^\times$ is open in $K^\times$, we further have $U_K^n \subseteq N_{L|K}L^\times$ for an appropriate n, hence $N_{L|K}L^\times$ contains the group $U_K^n \times (\pi)$.

(7.18) Corollary. *Every totally ramified abelian field $L|K$ is contained in some $L_{\pi,n}$.*

In view of Theorem (6.4) we also note the following fact:

(7.19) Theorem. *The group $U_K^n \times (\pi^f)$ is the norm group of the field $K' \cdot L_{\pi,n}$, where $K'|K$ is the unramified extension of degree f.*

Obviously $U_K^n \times (\pi^f) = (U_K \times (\pi^f)) \cap (U_K^n \times (\pi)) = N_{K'|K} K'^\times \cap N_{L_{\pi,n}|K} L_{\pi,n}^\times = N_{K' \cdot L_{\pi,n}|K} (K' \cdot L_{\pi,n})^\times$.

(7.20) Definition. *Let $L|K$ be an abelian extension, and let n be the smallest integer ≥ 0 such that $U_K^n \subseteq N_{L|K} L^\times$. Then the ideal*

$$\mathfrak{f} = \mathfrak{p}_K^n$$

*is called the **conductor** of $L|K$ [16].*

The conductor of the extension $L_{\pi,n}|K$ is the ideal $\mathfrak{f} = \mathfrak{p}_K^n$. For the unramified extensions $L|K$, we have the

(7.21) Theorem. *An abelian extension $L|K$ is unramified if and only if it has conductor $\mathfrak{f} = 1$.*

The follows immediately from (4.9), $L|K$ is unramified if and only if the norm group $N_{L|K} L^\times$ has the form $U_K^0 \times (\pi^f)$, thus if and only if $\mathfrak{f} = \mathfrak{p}_K^0 = 1$.

The notion of a conductor is closely related to the discriminant and plays a role in global class field theory[17].

We end this section with a brief discussion of the universal norm residue symbol $(\ , K)$, which is characterized by the following theorem.

(7.22) Theorem. *Let π be a prime element of K, $f \in \xi_\pi$, $\Lambda_f = \bigcup_{n=1}^\infty \Lambda_{f,n}$, $L_\pi = \bigcup_{n=1}^\infty L_{\pi,n} = K(\Lambda_f)$, and $G_\pi = G_{L_\pi|K}$.*

The field $T \cdot L_\pi$ is (independent of π) the maximal abelian field over K. Thus

$$G_K^{ab} = G_{T|K} \times G_\pi.$$

If $a = u \cdot \pi^m \in K^\times$, $u \in U_K$, then the norm residue symbol (a, K) is given by

$$(a, K)|_T = \varphi^m, \quad (a, K)\lambda = (u^{-1})_f(\lambda) \quad \text{for } \lambda \in \Lambda_f.$$

Proof. By (6.4) the norm group of an abelian field $L|K$ is a group containing the group $U_K^n \times (\pi^f)$. By (7.19) L is a subfield of a field $K' \cdot L_{\pi,n}$, $K' \subseteq T$, hence a subfield of $T \cdot L_\pi$, the maximal abelian extension of K.

---

[16] Here we set $U_K^0 = U_K$.

[17] Cf. [3], [20].

Because $G_K^{\mathrm{ab}} = G_{T|K} \times G_\pi$, the norm residue symbol (a, K) is determined by the equations

$$(a, K)|_T = \varphi^m, \ (a, K)\lambda = (u^{-1})_f(\lambda) \text{ for } \lambda \in \Lambda_f,$$

which hold by (4.10) and (7.15).

Part III

Global Class Field Theory

J. Neukirch, *Class Field Theory*, DOI 10.1007/978-3-642-35437-3_3,
© Springer-Verlag Berlin Heidelberg 2013

§ 1. Number Theoretic Preliminaries

We assume the reader is familiar with the basic concepts and theorems of algebraic number theory for which we refer to the standard text books, for example, [6], [21], [30]. Nevertheless, in this section we briefly summarize the for us most important facts.

If K is a finite algebraic number field, we mean by the primes $\mathfrak{p}$ of K the classes of equivalent valuations of K, where we distinguish between the finite and infinite primes $\mathfrak{p}$. The finite primes $\mathfrak{p}$ are associated with the nonarchimedian valuations of K and correspond bijectively to the prime ideals of the field K, for which we use the same symbol $\mathfrak{p}$. For the infinite primes we have to distinguish further between the real and complex infinite primes. The real primes correspond bijectively to the different embeddings of K into the field $\mathbb{R}$ of real numbers, while the complex primes correspond bijectively to the pairs of complex conjugate embeddings of K into the field $\mathbb{C}$ of complex numbers; we observe that two conjugate embeddings of K into $\mathbb{C}$ produce the same valuation of K. We write $\mathfrak{p} \nmid \infty$ (resp. $\mathfrak{p} \mid \infty$) if $\mathfrak{p}$ is finite (resp. infinite).

If $\mathfrak{p}$ is a finite prime, we denote by $v_\mathfrak{p}$ the exponential valuation of K associated with $\mathfrak{p}$, normalized with smallest positive value 1. We get an additional normalized valuation if we associate with each prime $\mathfrak{p}$ its $\mathfrak{p}$-absolute value $|\ |_\mathfrak{p}$. This is done in the following way:

1) If $\mathfrak{p}$ is finite and p is the rational prime lying under $\mathfrak{p}$, then for $a \in K$, $a \neq 0$, let $|a|_\mathfrak{p} = \mathfrak{N}(\mathfrak{p})^{-v_\mathfrak{p}(a)} = p^{-f_\mathfrak{p} \cdot v_\mathfrak{p}(a)}$. Here $\mathfrak{N}(\mathfrak{p})$ denotes the absolute norm of the ideal $\mathfrak{p}$, thus the number $p^{f_\mathfrak{p}}$ of elements in the residue field of $\mathfrak{p}$; $f_\mathfrak{p}$ is the inertia degree, i.e., the degree of the residue field of $\mathfrak{p}$ over its prime field (cf. II, §3, p. 79).

2) If $\mathfrak{p}$ is real infinite and ι is the embedding of K into the field $\mathbb{R}$ associated with $\mathfrak{p}$, then we set $|a|_\mathfrak{p} = |\iota a|$, $a \in K$.

3) If $\mathfrak{p}$ is complex infinite and ι is one of the pair of conjugate embeddings of K into the field $\mathbb{C}$ of complex numbers associated with $\mathfrak{p}$, then we let $|a|_\mathfrak{p} = |\iota a|^2$, $a \in K$.

With this normalization of the (multiplicative) valuations of K we have $|a|_\mathfrak{p} = 1$ ($a \in K \smallsetminus \{0\}$) for almost all primes $\mathfrak{p}$, and the fundamental **product formula**

$$\prod_\mathfrak{p} |a|_\mathfrak{p} = 1, \quad a \in K^\times. \text{ [1]}$$

---

[1] Cf. [21], III, §20, p. 314. As always, $K^\times$ denotes the multiplicative group of the field K.

If S is a finite set of primes of the field K containing all infinite primes, then
$$K^S = \{a \in K^\times \mid v_\mathfrak{p}(a) = 0 \ (\text{i.e.}, |a|_\mathfrak{p} = 1) \text{ for all } \mathfrak{p} \notin S\}$$
is called the group of **S-units** of K. In particular, if $S = S_\infty$ is the set of all infinite primes of K, then K^{S_∞} is the usual unit group of K. We have the following generalization of Dirichlet's

(1.1) Unit Theorem[2]. *The group K^S is finitely generated, and its rank is equal to $|S| - 1$, where $|S|$ denotes the number of primes in S.*

We write J_K for the group of ideals of K, and $P_K \subseteq J_K$ for the group of principal ideals. The factor group J_K/P_K is called the **ideal class group** of K. It satisfies the

(1.2) Theorem[3]. *The ideal class group J_K/P_K is finite; its order h is called the **class number** of the field K.*

For every prime $\mathfrak{p}$ of K we have the completion $K_\mathfrak{p}$ of K with respect to the valuation associated with $\mathfrak{p}$. If $\mathfrak{p}$ is finite, then $K_\mathfrak{p}$ is a $\mathfrak{p}$-adic number field. The prime $\mathfrak{p}$ is real (resp. complex) infinite if and only if $K_\mathfrak{p} = \mathbb{R}$, (resp. $K_\mathfrak{p} = \mathbb{C}$). We set
$$U_\mathfrak{p} = \begin{cases} \text{unit group of the field } K_\mathfrak{p}, \text{ if } \mathfrak{p} \text{ is finite;} \\ K_\mathfrak{p}^\times, \text{ if } \mathfrak{p} \text{ is infinite.} \end{cases}$$

It is convenient to define the unit groups $U_\mathfrak{p}$ for infinite primes as well, since we do not want to always have to distinguish between finite and infinite primes.

If $L|K$ is a finite extension of a number field K, we denote the primes in the extension field L by $\mathfrak{P}$. If $\mathfrak{P}$ is a prime of L lying above the prime $\mathfrak{p}$ of K, then we write $\mathfrak{P} \mid \mathfrak{p}$ for short. In this case the completion $L_\mathfrak{P}$ of L by $\mathfrak{P}$ contains the field $K_\mathfrak{p}$, since the restriction of the valuation associated with $\mathfrak{P}$ from L to K yields the valuation of the field K associated with $\mathfrak{p}$.

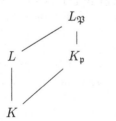

We have illustrated this situation in the adjacent diagram. The transition from the "global" extension $L|K$ to the "local" extensions $L_\mathfrak{P}|K_\mathfrak{p}$ at the individual primes is the fundamental principle behind class field theory.

If $\mathfrak{p}$ is a prime ideal of K and $\mathfrak{p} = \mathfrak{P}^e \cdots \mathfrak{P}'^{e'}$ is the prime decomposition of $\mathfrak{p}$ in the extension field L, then $\hat{\mathfrak{p}} = \hat{\mathfrak{P}}^e$, where $\hat{\mathfrak{p}}$ (resp. $\hat{\mathfrak{P}}$), denotes the prime

[2] Cf. [21], III, §28, p. 528.

[3] Cf. [21], III. §29, p. 542.

ideal of the field $K_{\mathfrak{p}}$ (resp. of the extension field $L_{\mathfrak{P}}$). Moreover, $\widehat{\mathfrak{P}}$ has the same degree over $\widehat{\mathfrak{p}}$ as $\mathfrak{P}$ over $\mathfrak{p}$. If $\mathfrak{P}$ runs through all the primes of L lying over $\mathfrak{p}$, then we have the fundamental equation of number theory

$$\sum_{\mathfrak{P}|\mathfrak{p}} [L_{\mathfrak{P}} : K_{\mathfrak{p}}] = [L : K].$$

Let $L|K$ be a finite normal field extension with Galois group $G = G_{L|K}$. If $\sigma \in G$, then along with $\mathfrak{P} \mid \mathfrak{p}$ we also have $\sigma\mathfrak{P} \mid \mathfrak{p}$, where $\sigma\mathfrak{P}$ is the prime of L conjugate to $\mathfrak{P}$ with respect to σ.

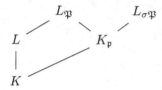

If we complete L with respect to $\mathfrak{P}$ and $\sigma\mathfrak{P}$, then $K_{\mathfrak{p}}$ is contained in $L_{\mathfrak{P}}$ as well as in $L_{\sigma\mathfrak{P}}$, since $\mathfrak{p}$ lies under $\mathfrak{P}$ as well as under $\sigma\mathfrak{P}$.

There is a canonical $K_{\mathfrak{p}}$-isomorphism

$$L_{\mathfrak{P}} \xrightarrow{\sigma} L_{\sigma\mathfrak{P}}$$

between $L_{\mathfrak{P}}$ and $L_{\sigma\mathfrak{P}}$ which we also denote by σ. In fact, if $\alpha \in L_{\mathfrak{P}}$, thus $\alpha = \mathfrak{P}\text{-}\lim \alpha_i$ for some sequence $\alpha_i \in L$, then the sequence $\sigma\alpha_i \in L$ converges in $L_{\sigma\mathfrak{P}}$ with respect to $\sigma\mathfrak{P}$, and the canonical isomorphism is obtained from

$$\alpha = \mathfrak{P}\text{-}\lim \alpha_i \in L_{\mathfrak{P}} \longmapsto \sigma\alpha = \sigma\mathfrak{P}\text{-}\lim \sigma\alpha_i \in L_{\sigma\mathfrak{P}}.$$

Under this isomorphism the field $K_{\mathfrak{p}}$ is obviously fixed elementwise. In particular, in case $\mathfrak{P} = \sigma\mathfrak{P}$ we obtain a $K_{\mathfrak{p}}$-automorphism

$$L_{\mathfrak{P}} \xrightarrow{\sigma} L_{\mathfrak{P}},$$

and therefore an element of the Galois group $G_{L_{\mathfrak{P}}|K_{\mathfrak{p}}}$ of $L_{\mathfrak{P}}|K_{\mathfrak{p}}$. This automorphism is simply the continuous extension of the automorphism σ of L to the completion $L_{\mathfrak{P}}$. If we observe that $\mathfrak{P} = \sigma\mathfrak{P}$ if and only if σ is an element of the decomposition group $G_{\mathfrak{P}} \subseteq G$ of $\mathfrak{P}$ over K, we see that for each $\sigma \in G_{\mathfrak{P}}$ there is a corresponding element in $G_{L_{\mathfrak{P}}|K_{\mathfrak{p}}}$. Conversely, every automorphism of $G_{L_{\mathfrak{P}}|K_{\mathfrak{p}}}$ yields an automorphism in $G_{\mathfrak{P}}$ by restriction to the field L. This yields a canonical isomorphism between the Galois group $G_{L_{\mathfrak{P}}|K_{\mathfrak{p}}}$ of the local extension $L_{\mathfrak{P}}|K_{\mathfrak{p}}$ and the decomposition group $G_{\mathfrak{P}}$, and we can identify $G_{L_{\mathfrak{P}}|K_{\mathfrak{p}}}$ and $G_{\mathfrak{P}}$; therefore we can consider $G_{L_{\mathfrak{P}}|K_{\mathfrak{p}}}$ as a subgroup of G, i.e., $G_{L_{\mathfrak{P}}|K_{\mathfrak{p}}} = G_{\mathfrak{P}} \subseteq G$. In what follows we will identify these groups without further mentioning.

The Theory of Kummer Extensions. Later we will apply the following observations to algebraic number fields; however, since they hold for arbitrary fields, we state them in full generality.

Let K be a field containing the n-th roots of unity but whose characteristic does not divide n. By a general **Kummer extension** of K we mean a (finite or infinite) Galois extension L of K whose Galois group $G_{L|K}$ is abelian and has exponent n, i.e., has the property that $\sigma^n = 1$ for all $\sigma \in G_{L|K}$. It is easy to see that the compositum of two Kummer extensions L_1 and L_2 of K is again a Kummer extension. Therefore the union N of all abelian extensions of K with a Galois group of exponent n is the largest Kummer extension of K. The following theorem shows that one can read off the structure of the Kummer extensions of K from the multiplicative group $K^\times$ of K.

(1.3) Theorem. *There exists an inclusion preserving isomorphism between the lattice of Kummer extensions L of K and the lattice of certain subgroups Δ of $K^\times$. Under this isomorphism the group $\Delta = (L^\times)^n \cap K^\times \supseteq (K^\times)^n$ corresponds to the Kummer extension L, and the field $L = K(\sqrt[n]{\Delta})$ corresponds to the group Δ, $(K^\times)^n \subseteq \Delta \subseteq K^\times$:*

$$L \longmapsto \Delta = (L^\times)^n \cap K^\times,$$
$$\Delta \longmapsto L = K(\sqrt[n]{\Delta}).$$

The factor group $\Delta/(K^\times)^n$ is isomorphic to the character group $\chi(G_{L|K})$ of the Galois group $G_{L|K}$.

Proof. Let L be a Kummer extension of K and $G = G_{L|K}$ its Galois group. Then G has exponent n and acts trivially on the group μ_n of n-th roots of unity which lies in K. Hence $\chi(G) = \mathrm{Hom}(G, \mu_n) = H^1(G, \mu_n)$. If we associate with every $x \in L^\times$ its n-th power $x^n \in (L^\times)^n$, we obtain the exact sequence

$$1 \longrightarrow \mu_n \longrightarrow L^\times \overset{n}{\longrightarrow} (L^\times)^n \longrightarrow 1$$

and from the resulting cohomology sequence the connecting homomorphism

$$L^{\times G} = K^\times \overset{n}{\to} ((L^\times)^n)^G = (L^\times)^n \cap K^\times \overset{\delta}{\to} H^1(G, \mu_n) \to H^1(G, L^\times) = 1 \;^{4)},$$

which yields the isomorphism

$$((L^\times)^n \cap K^\times)/(K^\times)^n \cong H^1(G, \mu_n) = \chi(G).$$

It is easy to check that this isomorphism takes a class $a \cdot (K^\times)^n \in ((L^\times)^n \cap K^\times)/(K^\times)^n$ to the character $\chi_a \in H^1(G, \mu_n)$ with $\chi_a(\sigma) = \sigma(\sqrt[n]{a})/\sqrt[n]{a}$.

In particular, if we consider the maximal Kummer extension N over K with the Galois group $G = G_{N|K}$, then $(N^\times)^n \cap K^\times = (K^\times)^n$, since if $(K^\times)^n \subset (N^\times)^n \cap K^\times$, then there would exist an element $a \in K^\times$ whose n-th root $\sqrt[n]{a}$ would not lie in $N^\times$, and therefore would generate over N a still larger

---

$^{4)}$ Note that $H^1(G, L^\times) = 1$ by II, (2.2)). Of course, the exactness of this sequence can be shown without reference to cohomology, and one can develop the theory of Kummer extensions without using cohomological methods.

Kummer extension over K. Consequently we have an isomorphism

$$K^\times/(K^\times)^n \cong H^1(G, \mu_n) = \chi(G).$$

By Pontryagin Duality, there is an inclusion reversing lattice isomorphism between the lattice of closed subgroups of G and the lattice of subgroups of $\chi(G)$. Since $\chi(G) \cong K^\times/(K^\times)^n$, we have by Galois theory an inclusion preserving isomorphism between the lattice of Kummer extensions and the lattice of subgroups of $K^\times$ which contain $(K^\times)^n$. If the field L corresponds to the group Δ in this sense, then

1. $\Delta = (L^\times)^n \cap K^\times$: If $a \in K^\times$, then $a \in \Delta \Leftrightarrow \chi_a(\sigma) = 1 \ (\chi_a \in \chi(G_{N|K}))$ for $\sigma \in G_{N|L} \Leftrightarrow \sigma(\sqrt[n]{a}) = \sqrt[n]{a}$ for $\sigma \in G_{N|L} \Leftrightarrow \sqrt[n]{a} \in L \Leftrightarrow a \in (L^\times)^n$.

2. $L = K(\sqrt[n]{\Delta})$: If $\sigma \in G = G_{N|K}$, then $\sigma \in G_{N|L} \Leftrightarrow \chi_a(\sigma) = 1$ for $a \in \Delta \Leftrightarrow \sigma(\sqrt[n]{a}) = \sqrt[n]{a}$ for $a \in \Delta \Leftrightarrow \sigma|_{K(\sqrt[n]{\Delta})} = \mathrm{id}_{K(\sqrt[n]{\Delta})} \Leftrightarrow \sigma \in G_{N|K(\sqrt[n]{\Delta})}$.

§ 2. Idèles and Idèle Classes

In the following we will consider **idèles**[5] which were first introduced by C. CHEVALLEY. The notion of idèles is a slight modification of the notion of ideals, or, more precisely, of divisors. Its significance lies in the fact that it permits a transition between global and local number theory, and therefore represents a suitable mean for applying the local-global principle, which is a method to obtain theorems and definitions in global class field theory from local class field theory. The development of the global theory using idèles together with cohomological methods is particularly transparent, and has led to a plethora of far-reaching results. As a result, the analytic methods, i.e., Dirichlet series and their generalizations, which were necessary in the classical ideal theoretic treatment of class field theory, have disappeared[6].

Let K be an algebraic number field. An **idèle** $\mathfrak{a}$ of K is a family $\mathfrak{a} = (\mathfrak{a}_\mathfrak{p})$ of elements $\mathfrak{a}_\mathfrak{p} \in K_\mathfrak{p}^\times$ such that $\mathfrak{p}$ ranges over all primes of K, but $\mathfrak{a}_\mathfrak{p}$ is a unit in $K_\mathfrak{p}$ for almost all primes $\mathfrak{p}$. We also obtain these idèles by the following

(2.1) Definition. *Let S be a finite set of primes of K. The group*

$$I_K^S = \prod_{\mathfrak{p} \in S} K_\mathfrak{p}^\times \times \prod_{\mathfrak{p} \notin S} U_\mathfrak{p} \subseteq \prod_\mathfrak{p} K_\mathfrak{p}^\times$$

---

[5] The idèles were first called **ideal elements**. Then this was abbreviated to id. el. from which the name idel (in French idèle) evolved.

[6] Yet even today these methods represent an essential counterpart to the approach we follow here, and have not lost their importance.

is called **the group of S-idèles** of K. The union

$$I_K = \bigcup_S I_K^S \subseteq \prod_{\mathfrak{p}} K_{\mathfrak{p}}^{\times},$$

where S runs through all finite sets of primes of K, is the **idèle group** of K. If $\mathfrak{a} = (\mathfrak{a}_{\mathfrak{p}}) \in I_K$, $\mathfrak{a}_{\mathfrak{p}} \in K_{\mathfrak{p}}^{\times}$, then the $\mathfrak{a}_{\mathfrak{p}}$ are the **local components** of the idèle $\mathfrak{a}$; an $\mathfrak{a}_{\mathfrak{p}} \in K_{\mathfrak{p}}^{\times}$ is an **essential component** of $\mathfrak{a}$ if $\mathfrak{a}_{\mathfrak{p}}$ is not a unit.

In particular, an idèle has at most finitely many essential components. The S-idèles are precisely those idèles which have essential components at most at the primes in the set S.

The reason why we allow units as "nonessential" components at almost all primes of an idèle is that this allows us to canonically embed the multiplicative group $K^{\times}$ of the field K into the idèle group I_K of K:

If $x \in K^{\times}$, then we let $(x) \in I_K$ be the idèle whose components are $(x)_{\mathfrak{p}} = x \in K_{\mathfrak{p}}^{\times}$. Observe that x is a unit in $K_{\mathfrak{p}}$ for almost all primes $\mathfrak{p}$. In this way we always consider $K^{\times}$ as embedded in I_K, thus view $K^{\times}$ as a subgroup of I_K. The idèles in $K^{\times}$ are called the **principal idèles** of K.

If S is a finite set of primes of K, then we denote by

$$K^S = K^{\times} \cap I_K^S \subseteq I_K^S$$

the group of S-principal idèles. The elements in K^S are also called the **S-units** of K, since they are units for all primes $\mathfrak{p} \notin S$ (cf. §1, p. 114). In particular, if $S = S_{\infty}$ is the set of all infinite primes of K, then $K^{S_{\infty}}$ is the ordinary unit group of the field K.

(2.2) Definition. *The factor group*

$$C_K = I_K / K^{\times}$$

*is called the **idèle class group** of the field K.*

In our development of class field theory the group C_K is the main object of interest. The connection between the idèles and the ideals of a field K is given in the following proposition.

(2.3) Proposition. *Let S_{∞} be the set of all infinite primes of the field K, and let $I_K^{S_{\infty}}$ be the group of idèles which have units as components at all finite primes. Then we have a canonical isomorphism*

$$I_K / I_K^{S_{\infty}} \cong J_K, \quad I_K / I_K^{S_{\infty}} \cdot K^{\times} \cong J_K / P_K,$$

where J_K and P_K denote the group of ideals and principal ideals respectively.

Proof. If $\mathfrak{p}$ is a finite prime of K, we let $v_\mathfrak{p}$ be the valuation of $K_\mathfrak{p}$ normalized with minimal positive value 1. If $\mathfrak{a} \in I_K$ is an idèle of K, then we have $\mathfrak{a}_\mathfrak{p} \in U_\mathfrak{p}$ for almost all finite primes, and therefore $v_\mathfrak{p}\mathfrak{a}_\mathfrak{p} = 0$. We obtain from the map

$$\mathfrak{a} \longmapsto \prod_{\mathfrak{p} \nmid \infty} \mathfrak{p}^{v_\mathfrak{p}\mathfrak{a}_\mathfrak{p}} ,$$

where $\mathfrak{p}$ ranges over all finite primes of K, a canonical homomorphism from I_K onto J_K. Here $\mathfrak{a}$ is mapped to the unit ideal if and only if $v_\mathfrak{p}\mathfrak{a}_\mathfrak{p} = 0$, i.e., $\mathfrak{a}_\mathfrak{p} \in U_\mathfrak{p}$ for all $\mathfrak{p} \nmid \infty$, thus if and only if $\mathfrak{a} \in I_K^{S_\infty}$. Hence the kernel of this homomorphism is $I_K^{S_\infty}$. On the other hand, we have the homomorphism

$$\mathfrak{a} \longmapsto \prod_{\mathfrak{p} \nmid \infty} \mathfrak{p}^{v_\mathfrak{p}\mathfrak{a}_\mathfrak{p}} \cdot P_K$$

from I_K onto J_K/P_K, and $\mathfrak{a}$ is in its kernel if and only if $\prod_{\mathfrak{p} \nmid \infty} \mathfrak{p}^{v_\mathfrak{p}\mathfrak{a}_\mathfrak{p}} \in P_K$, i.e., $\prod_{\mathfrak{p} \nmid \infty} \mathfrak{p}^{v_\mathfrak{p}\mathfrak{a}_\mathfrak{p}} = (x) = \prod_{\mathfrak{p} \nmid \infty} \mathfrak{p}^{v_\mathfrak{p}x}$ with $x \in K^\times$, thus if and only if $v_\mathfrak{p}\mathfrak{a}_\mathfrak{p} = v_\mathfrak{p}x$, $v_\mathfrak{p}(\mathfrak{a}_\mathfrak{p} \cdot x^{-1}) = 0$ for all $\mathfrak{p} \nmid \infty$. This again is the case if and only if $\mathfrak{a} \cdot x^{-1} \in I_K^{S_\infty}$, thus $\mathfrak{a} \in x \cdot I_K^{S_\infty}$, i.e., if and only if $\mathfrak{a} \in I_K^{S_\infty} \cdot K^\times$. $\qquad \square$

The group $I_K/I_K^{S_\infty}$ is none other than the well-known group of fractional ideals of K. It is easy to see that when passing to ideals, the S-idèles become precisely those ideals that are generated only by prime ideals in S.

Unlike the ideal class group J_K/P_K, the idèle class group $C_K = I_K/K^\times$ is not finite. However, the finiteness of the ideal class group is reflected in the fact that all idèle classes in C_K can be represented by S-idèles $\mathfrak{a} \in I_K^S$ for a fixed finite set S of primes. This is the assertion of the following proposition.

(2.4) Proposition. *Let S be a sufficiently large finite set of primes. Then*
$$I_K = I_K^S \cdot K^\times, \quad \text{and therefore} \quad C_K = I_K^S \cdot K^\times / K^\times.$$

Proof. The ideal class group J_K/P_K is finite (cf. (1.2)). Hence we can choose a finite set of ideals $\mathfrak{A}_1, \dots, \mathfrak{A}_n$ which represent the classes in J_K/P_K. The ideals $\mathfrak{A}_1, \dots, \mathfrak{A}_n$ are further made up of only finitely many prime ideals $\mathfrak{p}_1, \dots, \mathfrak{p}_s$. Now if S is any finite set of primes containing the primes $\mathfrak{p}_1, \dots, \mathfrak{p}_s$ and all the infinite primes of K, then in fact

$$I_K = I_K^S \cdot K^\times.$$

In order to see this, consider the isomorphism $I_K/I_K^{S_\infty} \cong J_K$ (cf. (2.3)). If $\mathfrak{a} \in I_K$, then the corresponding ideal $\mathfrak{A} = \prod_{\mathfrak{p} \nmid \infty} \mathfrak{p}^{v_\mathfrak{p}\mathfrak{a}_\mathfrak{p}}$ lies in a class $\mathfrak{A}_i \cdot P_K$, i.e., $\mathfrak{A} = \mathfrak{A}_i \cdot (x)$, where $(x) \in P_K$ denotes the principal ideal given by $x \in K^\times$. The idèle $\mathfrak{a}' = \mathfrak{a} \cdot x^{-1}$ is mapped under the homomorphism $I_K \to J_K$ onto the ideal $\mathfrak{A}' = \prod_{\mathfrak{p} \nmid \infty} \mathfrak{p}^{v_\mathfrak{p}\mathfrak{a}'_\mathfrak{p}} = \mathfrak{A}_i$. Since the prime ideal components of $\mathfrak{A}_i$ lie in the set S, we have $v_{\mathfrak{p}} \mathfrak{a}'_{\mathfrak{p}} = 0$, i.e., $\mathfrak{a}'_{\mathfrak{p}} \in U_{\mathfrak{p}}$ for all $\mathfrak{p} \notin S$; thus $\mathfrak{a}' = \mathfrak{a} \cdot x^{-1} \in I_K^S$, $\mathfrak{a} \in I_K^S \cdot K^{\times}$.

We now investigate what happens when we pass to an extension field.

Let $L|K$ be a finite extension of algebraic number fields. If $\mathfrak{p}$ is a prime of K and $\mathfrak{P}$ a prime of L lying over $\mathfrak{p}$, we write simply $\mathfrak{P} \mid \mathfrak{p}$. The idèle group I_K of K is embedded in the idèle group I_L of L as follows: If we map an idèle $\mathfrak{a} \in I_K$ to the idèle $\mathfrak{a}' \in I_L$ with the components

$$\mathfrak{a}'_{\mathfrak{P}} = \mathfrak{a}_{\mathfrak{p}} \in K_{\mathfrak{p}} \subseteq L_{\mathfrak{P}} \qquad \text{for} \quad \mathfrak{P} \mid \mathfrak{p},$$

then we obtain an injective homomorphism

$$I_K \longrightarrow I_L .$$

This homomorphism allows us to think of I_K as being embedded in I_L, and to regard I_K as a subgroup of I_L. With this identification, an idèle $\mathfrak{a} \in I_L$ is in the group I_K if and only if its components $\mathfrak{a}_{\mathfrak{P}}$ lie in $K_{\mathfrak{p}}$ (where $\mathfrak{P} \mid \mathfrak{p}$), and moreover any two primes $\mathfrak{P}$ and $\mathfrak{P}'$ lying over the same prime $\mathfrak{p}$ of K have equal components $\mathfrak{a}_{\mathfrak{P}} = \mathfrak{a}_{\mathfrak{P}'} \in K_{\mathfrak{p}}$.

If $L|K$ is normal and $G = G_{L|K}$ denotes its Galois group, I_L is canonically a G-module: An element $\sigma \in G$ defines a canonical isomorphism from $L_{\sigma^{-1}\mathfrak{P}}$ onto $L_{\mathfrak{P}}$, which we also denote by σ (cf. §1, p. 115). Here we associate with an idèle $\mathfrak{a} \in I_L$ with components $\mathfrak{a}_{\mathfrak{P}} \in L_{\mathfrak{P}}^{\times}$ the idèle $\sigma \mathfrak{a} \in I_L$ with components

$$(\sigma \mathfrak{a})_{\mathfrak{P}} = \sigma \mathfrak{a}_{\sigma^{-1}\mathfrak{P}} \in L_{\mathfrak{P}} .$$

Note that $\mathfrak{a}_{\sigma^{-1}\mathfrak{P}} \in L_{\sigma^{-1}\mathfrak{P}}$ is the $\sigma^{-1}\mathfrak{P}$-component of $\mathfrak{a}$, which is mapped by σ into $L_{\mathfrak{P}}$. If we take into account that the $\mathfrak{P}$-component $(\sigma \mathfrak{a})_{\mathfrak{P}}$ of $\sigma \mathfrak{a}$ is essential if and only if the $\sigma^{-1}\mathfrak{P}$-component $\mathfrak{a}_{\sigma^{-1}\mathfrak{P}}$ of $\mathfrak{a}$ is essential, we immediately see that when passing to ideals, the map induced by $\mathfrak{a} \mapsto \sigma \mathfrak{a}$ is just the ordinary conjugation map on the ideal group J_L.

(2.5) Proposition. *Let $L|K$ be normal with Galois group $G = G_{L|K}$. Then*

$$I_L^G = I_K.$$

Proof. The inclusion $I_K \subseteq I_L^G$ is easy. If $\sigma \in G$, then the isomorphism $L_{\sigma^{-1}\mathfrak{P}} \xrightarrow{\sigma} L_{\mathfrak{P}}$ is a $K_{\mathfrak{p}}$-isomorphism ($\mathfrak{P} \mid \mathfrak{p}$), and if $\mathfrak{a} \in I_K$ is considered as an idèle of I_L, then $(\sigma \mathfrak{a})_{\mathfrak{P}} = \sigma \mathfrak{a}_{\sigma^{-1}\mathfrak{P}} = \sigma \mathfrak{a}_{\mathfrak{P}} = \mathfrak{a}_{\mathfrak{P}} \in K_{\mathfrak{p}}$, i.e., $\sigma \mathfrak{a} = \mathfrak{a}$.

For the inclusion $I_L^G \subseteq I_K$, consider $\mathfrak{a} \in I_L$ with $\sigma \mathfrak{a} = \mathfrak{a}$ for all $\sigma \in G$. Then $(\sigma \mathfrak{a})_{\mathfrak{P}} = \sigma \mathfrak{a}_{\sigma^{-1}\mathfrak{P}} = \mathfrak{a}_{\mathfrak{P}}$ for all primes $\mathfrak{P}$ of L. By §1, p. 115 we can consider the decomposition group $G_{\mathfrak{P}}$ of $\mathfrak{P}$ over K as the Galois group of the extension $L_{\mathfrak{P}}|K_{\mathfrak{p}}$. For every $\sigma \in G_{\mathfrak{P}}$ we have $\sigma^{-1}\mathfrak{P} = \mathfrak{P}$, and because $\mathfrak{a}_{\mathfrak{P}} = \sigma \mathfrak{a}_{\sigma^{-1}\mathfrak{P}} = \sigma \mathfrak{a}_{\mathfrak{P}}$, we obtain $\mathfrak{a}_{\mathfrak{P}} \in K_{\mathfrak{p}}$ ($\mathfrak{P} \mid \mathfrak{p}$). Hence if σ is an arbitrary element of G, then $(\sigma \mathfrak{a})_{\mathfrak{P}} = \mathfrak{a}_{\mathfrak{P}} = \sigma \mathfrak{a}_{\sigma^{-1}\mathfrak{P}} = \mathfrak{a}_{\sigma^{-1}\mathfrak{P}} \in K_{\mathfrak{p}}$, i.e., two primes

$\mathfrak{P}$ and $\sigma^{-1}\mathfrak{P}$ lying above the same prime $\mathfrak{p}$ of K have the same components $\mathfrak{a}_{\mathfrak{P}} = \mathfrak{a}_{\sigma^{-1}\mathfrak{P}} \in K_{\mathfrak{p}}$, so that $\mathfrak{a} \in I_K$.

It is well known that an ideal of a field K can very well become a principal ideal in an extension field L without being a principal ideal in the base field K. The following proposition shows that the idèles behave differently.

(2.6) Proposition. *If $L|K$ is an arbitrary finite extension, then*

$$L^\times \cap I_K = K^\times .$$

In particular, if $\mathfrak{a} \in I_K$ is an idèle of K that becomes a principal idèle in the extension L, i.e., $\mathfrak{a} \in L^\times$, then $\mathfrak{a}$ is already principal in K.

Proof. The inclusion $K^\times \subseteq L^\times \cap I_K$ is trivial. Let $\tilde{L}$ be a finite normal extension of K containing L, and let $\tilde{G} = G_{\tilde{L}|K}$ be its Galois group. Then I_K and I_L are subgroups of $I_{\tilde{L}}$. If $\mathfrak{a} \in \tilde{L}^\times \cap I_K$, then (2.5) shows that $\mathfrak{a} \in I_{\tilde{L}}^{\tilde{G}}$, i.e., $\sigma\mathfrak{a} = \mathfrak{a}$ for all $\sigma \in \tilde{G}$, and because $\mathfrak{a} \in \tilde{L}^\times$, we even have $\mathfrak{a} \in (\tilde{L}^\times)^{\tilde{G}} = K^\times$. Therefore $\tilde{L}^\times \cap I_K = K^\times$, which implies $L^\times \cap I_K \subseteq \tilde{L}^\times \cap I_K = K^\times$.

By (2.6) we can embed the idèle class group C_K of a field K into the idèle class group C_L of a finite extension field L using the canonical homomorphism

$$\iota : C_K \longrightarrow C_L, \quad \mathfrak{a} \cdot K^\times \mapsto \mathfrak{a} \cdot L^\times \qquad (\mathfrak{a} \in I_K \subseteq I_L).$$

To see that ι is injective note that if the class $\mathfrak{a} \cdot K^\times \in C_K$ is mapped to the unit class $L^\times \in C_L$, hence $\mathfrak{a} \cdot L^\times = L^\times$, $\mathfrak{a} \in L^\times$, then we know by (2.6) that $\mathfrak{a} \in L^\times \cap I_K = K^\times$, i.e., $\mathfrak{a} \cdot K^\times = K^\times$ is the unit class of C_K.

In the following we view C_K as embedded in C_L via this canonical map, hence as a subgroup of C_L. An element $\mathfrak{a} \cdot L^\times \in C_L$ $(\mathfrak{a} \in I_L)$ lies in C_K if and only if the class $\mathfrak{a} \cdot L^\times$ contains a representative $\mathfrak{a}'$ from I_K $(\subseteq I_L)$ such that $\mathfrak{a}' \cdot L^\times = \mathfrak{a} \cdot L^\times$.

(2.7) Theorem. *Let $L|K$ be with Galois group $G = G_{L|K}$. Then C_L is canonically a G-module, and*

$$C_L^G = C_K.$$

Proof. If $\mathfrak{a} \cdot L^\times \in C_L$ $(\mathfrak{a} \in I_L)$, we set $\sigma(\mathfrak{a} \cdot L^\times) = \sigma\mathfrak{a} \cdot L^\times$. This definition is clearly independent of the choice of $\mathfrak{a} \in I_L$, and makes C_L a G-module.

From the exact sequence of G-module $1 \to L^\times \to I_L \to C_L \to 1$ we obtain the exact cohomology sequence (cf. I, (3.4))

$$1 \longrightarrow (L^\times)^G \longrightarrow I_L^G \longrightarrow C_L^G \longrightarrow H^1(G, L^\times),$$

where $(L^\times)^G = K^\times$, $I_L^G = I_K$, and $H^1(G, L^\times) = 1$; hence $C_L^G = C_K$.

We briefly summarize the most important results from this section.

Let K be an algebraic number field. Then

$$I_K^S = \prod_{\mathfrak{p} \in S} K_{\mathfrak{p}}^\times \times \prod_{\mathfrak{p} \notin S} U_{\mathfrak{p}} \qquad \text{the group of } S\text{-idèles of } K$$

$\qquad\qquad\qquad\qquad\qquad\qquad\qquad\;\;$ (S a finite set of primes of K),

$$I_K = \bigcup_S I_K^S \qquad\qquad\qquad\qquad \text{the idèle group of } K,$$

$$K^\times \subseteq I_K \qquad\qquad\qquad\qquad\quad\; \text{the group of principal idèles,}$$

$$C_K = I_K/K^\times \qquad\qquad\qquad\quad \text{the idèle class group of } K,$$

$$I_K = I_K^S \cdot K^\times \qquad\qquad\qquad\quad \text{for a sufficiently large finite set } S$$

$\qquad\qquad\qquad\qquad\qquad\qquad\qquad\;\;$ of primes.

If $L|K$ is a finite extension of algebraic number fields, we have embeddings

$$I_K \subseteq I_L \qquad\qquad \text{idèle groups,}$$

$$K^\times \subseteq L^\times \qquad\quad \text{groups of principal idèles,}$$

$$C_K \subseteq C_L \qquad\qquad \text{idèle class groups.}$$

If $L|K$ is a finite normal extension with Galois group $G = G_{L|K}$, then $L^\times$, I_L and C_L are G-modules whose G-invariants are given by

$$L^{\times G} = K^\times, \quad I_L^G = I_K, \quad C_L^G = C_K .$$

§3. Cohomology of the Idèle Group

Let $L|K$ be a finite normal extension with Galois group $G = G_{L|K}$. We consider the cohomology groups $H^q(G, I_L)$ of the G-module I_L. These cohomology groups reveal a particular advantage of working with idèles since they can in a certain sense be completely "localized", i.e., decomposed into a direct product of cohomology groups over the local fields $K_{\mathfrak{p}}$. The goal of this section is to explain this natural localization process.

Let S be a finite set of primes of the base field K and $\bar{S}$ the finite set of primes of the extension field L above the primes in S. To simplify things, we denote the group of $\bar{S}$-idèles $I_L^{\bar{S}}$ of L also by I_L^S, and speak of the S-idèles of the field L; we will use the same convention in later sections as well. Thus we have

$$I_L^S = \prod_{\mathfrak{P}|\mathfrak{p} \in S} L_{\mathfrak{P}}^{\times} \times \prod_{\mathfrak{P}|\mathfrak{p} \notin S} U_{\mathfrak{P}} = \prod_{\mathfrak{p} \in S} \prod_{\mathfrak{P}|\mathfrak{p}} L_{\mathfrak{P}}^{\times} \times \prod_{\mathfrak{p} \notin S} \prod_{\mathfrak{P}|\mathfrak{p}} U_{\mathfrak{P}}.$$

We consider the products $I_L^{\mathfrak{p}} = \prod_{\mathfrak{P}|\mathfrak{p}} L_{\mathfrak{P}}^{\times}$ and $U_L^{\mathfrak{p}} = \prod_{\mathfrak{P}|\mathfrak{p}} U_{\mathfrak{P}}$ as subgroups of I_L^S, where we think of the elements in $I_L^{\mathfrak{p}}$ (resp. in $U_L^{\mathfrak{p}}$), as those idèles which have the component 1 at all the primes of L not lying above $\mathfrak{p}$ (resp. and in addition have only units as components at the primes of L lying above $\mathfrak{p}$). Since the automorphisms $\sigma \in G$ only permute the primes $\mathfrak{P}$ above $\mathfrak{p}$, the groups $I_L^{\mathfrak{p}}$ and $U_L^{\mathfrak{p}}$ are G-modules. Thus we have decomposed I_L^S into a direct product of G-modules:

$$I_L^S = \prod_{\mathfrak{p} \in S} I_L^{\mathfrak{p}} \times \prod_{\mathfrak{p} \notin S} U_L^{\mathfrak{p}}.$$

About the G-modules $I_L^{\mathfrak{p}}$ and $U_L^{\mathfrak{p}}$ we have the following

(3.1) Proposition. *Let $\mathfrak{P}$ is a prime of L lying over $\mathfrak{p}$. Then*

$$H^q(G, I_L^{\mathfrak{p}}) \cong H^q(G_{\mathfrak{P}}, L_{\mathfrak{P}}^{\times}),$$

where $G_{\mathfrak{P}}$ is the decomposition group of $\mathfrak{P}$ over K, considered also as the Galois group of $L_{\mathfrak{P}}|K_{\mathfrak{p}}$. If $\mathfrak{p}$ is a finite unramified prime in L, then for all q

$$H^q(G, U_L^{\mathfrak{p}}) = 1.$$

Addendum. *The above isomorphism above is given by the composition*

$$H^q(G, I_L^{\mathfrak{p}}) \xrightarrow{\mathrm{res}} H^q(G_{\mathfrak{P}}, I_L^{\mathfrak{p}}) \xrightarrow{\bar{\pi}} H^q(G_{\mathfrak{P}}, L_{\mathfrak{P}}^{\times}),$$

where $\bar{\pi}$ is induced by the canonical projection $I_L^{\mathfrak{p}} \xrightarrow{\pi} L_{\mathfrak{P}}^{\times}$ that takes each idèle in $I_L^{\mathfrak{p}}$ to its $\mathfrak{P}$-component.

Proof. If $\sigma \in G$ runs through a system of representatives of the cosets $G/G_{\mathfrak{P}}$, we write for simplicity $\sigma \in G/G_{\mathfrak{P}}$, then $\sigma\mathfrak{P}$ runs through all distinct primes of L above $\mathfrak{p}$. Hence

$$I_L^{\mathfrak{p}} = \prod_{\sigma \in G/G_{\mathfrak{P}}} L_{\sigma\mathfrak{P}}^{\times} = \prod_{\sigma \in G/G_{\mathfrak{P}}} \sigma L_{\mathfrak{P}}^{\times} \quad \text{and} \quad U_L^{\mathfrak{p}} = \prod_{\sigma \in G/G_{\mathfrak{P}}} U_{\sigma\mathfrak{P}} = \prod_{\sigma \in G/G_{\mathfrak{P}}} \sigma U_{\mathfrak{P}},$$

which shows that $I_L^{\mathfrak{p}}$ and $U_L^{\mathfrak{p}}$ are $G/G_{\mathfrak{P}}$-induced G-modules. Applying Shapiro's Lemma I, (4.19) yields

$$H^q(G, I_L^{\mathfrak{p}}) \cong H^q(G_{\mathfrak{P}}, L_{\mathfrak{P}}^{\times}) \quad \text{and} \quad H^q(G, U_L^{\mathfrak{p}}) \cong H^q(G_{\mathfrak{P}}, U_{\mathfrak{P}}),$$

where the isomorphism $H^q(G, I_L^{\mathfrak{p}}) \to H^q(G_{\mathfrak{P}}, L_{\mathfrak{P}}^{\times})$ is the composition of the homomorphisms res and $\bar{\pi}$ described in the addendum. If $\mathfrak{p}$ is unramified in L, then the extension $L_{\mathfrak{P}}|K_{\mathfrak{p}}$ is unramified, and we can refer to local class field theory (cf. II, (4.3)) to obtain the result $H^q(G, U_L^{\mathfrak{p}}) \cong H^q(G_{\mathfrak{P}}, U_{\mathfrak{P}}) = 1$.

Because of Proposition (3.1) and the decomposition $I_L^S = \prod_{\mathfrak{p} \in S} I_L^{\mathfrak{p}} \times \prod_{\mathfrak{p} \notin S} U_L^{\mathfrak{p}}$ the cohomology groups of the idèle groups I_L^S and I_L are easy to compute. By I, (3.8) we have

$$H^q(G, I_L^S) \cong \prod_{\mathfrak{p} \in S} H^q(G, I_L^{\mathfrak{p}}) \times \prod_{\mathfrak{p} \notin S} H^q(G, U_L^{\mathfrak{p}}).$$

If the finite set S contains all (finite) primes of K which are ramified in L, then by (3.1) $H^q(G, I_L^{\mathfrak{p}}) = H^q(G_{\mathfrak{P}}, L_{\mathfrak{P}}^{\times})$ ($\mathfrak{P}$ any prime above $\mathfrak{p}$), and $H^q(G, U_L^{\mathfrak{p}}) = 1$ for each $\mathfrak{p} \notin S$. Therefore

$$H^q(G, I_L^S) \cong \prod_{\mathfrak{p} \in S} H^q(G_{\mathfrak{P}}, L_{\mathfrak{P}}^{\times}),$$

where $\mathfrak{P}$ denotes any prime above $\mathfrak{p}$.

Because $I_L = \bigcup_S I_L^S$, we also have

$$H^q(G, I_L) \cong \varinjlim_S H^q(G, I_L^S) \cong \varinjlim_S \prod_{\mathfrak{p} \in S} H^q(G_{\mathfrak{P}}, L_{\mathfrak{P}}^{\times}) \cong \bigoplus_{\mathfrak{p}} H^q(G_{\mathfrak{P}}, L_{\mathfrak{P}}^{\times}) \quad {}^{[7]},$$

where S runs through all finite sets of primes of K which contain all ramified primes [8]. Thus we have proved the following theorem:

(3.2) Theorem. *Let S be a finite set of primes of K which contains all primes ramified in L. Then*

$$H^q(G, I_L^S) \cong \prod_{\mathfrak{p} \in S} H^q(G_{\mathfrak{P}}, L_{\mathfrak{P}}^{\times}),$$

$$H^q(G, I_L) \cong \bigoplus_{\mathfrak{p}} H^q(G_{\mathfrak{P}}, L_{\mathfrak{P}}^{\times}). \qquad {}^{[7]}$$

Here $\mathfrak{P}$ denotes any prime above $\mathfrak{p}$.

From the proof and the Addendum (3.1) we obtain further the

Addendum. *The isomorphism $H^q(G, I_L) \cong \bigoplus_{\mathfrak{p}} H^q(G_{\mathfrak{P}}, L_{\mathfrak{P}}^{\times})$ is given by the projections $H^q(G, I_L) \longrightarrow H^q(G_{\mathfrak{P}}, L_{\mathfrak{P}}^{\times})$, i.e., the composition of the maps*

$$H^q(G, I_L) \xrightarrow{\mathrm{res}} H^q(G_{\mathfrak{P}}, I_L) \xrightarrow{\bar{\pi}} H^q(G_{\mathfrak{P}}, L_{\mathfrak{P}}^{\times}),$$

where $\bar{\pi}$ is induced by the canonical projection $I_L \xrightarrow{\pi} L_{\mathfrak{P}}^{\times}$ which takes each idèle $\mathfrak{a}$ to its $\mathfrak{P}$-component $\mathfrak{a}_{\mathfrak{P}}$.

---

[7] By the symbol $\bigoplus$ we mean the **direct sum**, i.e., the (here multiplicative) group of families $(\ldots, c_{\mathfrak{p}}, \ldots)$, in which only finitely components $c_{\mathfrak{p}}$ not equal to 1 appear. By contrast $\prod$ means the **direct product**, i.e., the group of all families $(\ldots, c_{\mathfrak{p}}, \ldots)$.

[8] One can prove $H^q(G, I_L) \cong \bigoplus_{\mathfrak{p}} H^q(G_{\mathfrak{P}}, L_{\mathfrak{P}}^{\times})$ directly without forming the somewhat mysterious limit: View an element of $H^q(G, I_L)$ as an element of $H^q(G, I_L^S)$ for an appropriate S, and map it by the isomorphism $H^q(G, I_L^S) \cong \prod_{\mathfrak{p} \in S} H^q(G_{\mathfrak{P}}, L_{\mathfrak{P}}^{\times})$ into the group $\prod_{\mathfrak{p} \in S} H^q(G_{\mathfrak{P}}, L_{\mathfrak{P}}^{\times})$, which also can be viewed as a subgroup of $\bigoplus_{\mathfrak{p}} H^q(G_{\mathfrak{P}}, L_{\mathfrak{P}}^{\times})$. It is easy to verify that this is an isomorphism.

The above projections map each element $c \in H^q(G, I_L)$ to its $\mathfrak{p}$-components $c_\mathfrak{p} \in H^q(G_\mathfrak{P}, L_\mathfrak{P}^\times)$ (as always, $\mathfrak{P}$ is a fixed prime above $\mathfrak{p}$). The theorem says that each c is uniquely determined by its **local components** $c_\mathfrak{p}$, which because of the direct sum $\bigoplus$ are almost all equal to 1. In dimensions $q > 0$ the map $c \mapsto c_\mathfrak{p}$ can be described in the following simple way. Given a cohomology class $c \in H^q(G, I_L)$, choose a cocycle $\mathfrak{a}(\sigma_1, \ldots, \sigma_q)$ representing c. This is a function on the group G with values in the idèle group I_L. Restrict this function to the group $G_\mathfrak{P}$, and take the $\mathfrak{P}$-components $\mathfrak{a}_\mathfrak{P}(\sigma_1, \ldots, \sigma_q)$ of the idèle $\mathfrak{a}(\sigma_1, \ldots, \sigma_q)$. The resulting function from $G_\mathfrak{P}$ to $L_\mathfrak{P}^\times$ is a cocyle, and its cohomology class $c_\mathfrak{p} \in H^q(G_\mathfrak{P}, L_\mathfrak{P}^\times)$ is the $\mathfrak{p}$-component of c.

The following proposition shows how changing the field affects taking local components.

(3.3) Proposition. *Let $N \supseteq L \supseteq K$ be normal extensions of K, and let $\mathfrak{P}' \mid \mathfrak{P} \mid \mathfrak{p}$ be primes of N, L, and K respectively. Then*

$$
\begin{aligned}
(\mathrm{inf}_N c)_\mathfrak{p} &= \mathrm{inf}_{N_{\mathfrak{P}'}}(c_\mathfrak{p}), & c &\in H^q(G_{L|K}, I_L), \quad q \geq 1, \\
(\mathrm{res}_L c)_\mathfrak{P} &= \mathrm{res}_{L_\mathfrak{P}}(c_\mathfrak{p}), & c &\in H^q(G_{N|K}, I_N), \\
(\mathrm{cor}_K c)_\mathfrak{p} &= \textstyle\sum_{\mathfrak{P}|\mathfrak{p}} \mathrm{cor}_{K_\mathfrak{p}}(c_\mathfrak{P}), & c &\in H^q(G_{N|L}, I_N).
\end{aligned}
$$

For the last two formulas it suffices to assume that only $N|K$ is normal.

For the third formula note that for each $\mathfrak{P} \mid \mathfrak{p}$ we choose a prime $\mathfrak{P}'$ of N lying above $\mathfrak{P}$, thus the corestrictions $\mathrm{cor}_{K_\mathfrak{p}}(c_\mathfrak{P})$ lie in (a priori) distinct cohomology groups $H^q(G_{N_{\mathfrak{P}'}|K_\mathfrak{p}}, N_{\mathfrak{P}'}^\times)$. However, we can identify these as follows: Given two primes of N lying over $\mathfrak{p}$, there is an automorphism $\sigma \in G_{N|K}$ that interchanges these primes; given this, the isomorphism $N_{\mathfrak{P}'}^\times \xrightarrow{\sigma} N_{\sigma\mathfrak{P}'}^\times$ induces a canonical isomorphism $H^q(G_{N_{\mathfrak{P}'}|K_\mathfrak{p}}, N_{\mathfrak{P}'}^\times) \cong H^q(G_{N_{\sigma\mathfrak{P}'}|K_\mathfrak{p}}, N_{\sigma\mathfrak{P}'}^\times)$ (cf. also (3.1)). Hence we may view $\mathrm{cor}_{K_\mathfrak{p}}(c_\mathfrak{P})$ for each $\mathfrak{P} \mid \mathfrak{p}$ as an element of the group $H^q(G_{N_{\mathfrak{P}'}|K_\mathfrak{p}}, N_{\mathfrak{P}'}^\times)$ for a fixed choice of $\mathfrak{P}' \mid \mathfrak{p}$, and form the sum in this group.

The proof of Proposition (3.3) uses the general and purely cohomological fact that the restriction map which occurs when passing to the local components commutes with the maps inf, res, and cor. This is easy to see at the cocycle level for inf and res if $q \geq 1$, and for cor if $q = -1, 0$. The general case follows from this by dimension shifting. The details are left to the reader.

With Theorem (3.2) we have achieved a complete "localization" of the cohomology of the idèle groups, i.e., instead of the groups $H^q(G, I_L)$ we may consider the local cohomology groups $H^q(G_\mathfrak{P}, L_\mathfrak{P}^\times)$, a principle, which we will frequently use below. We make the important remark that Theorem (3.2) is not a deep number-theoretic result. Apart from purely cohomological results, the only number-theoretical facts that are being used in its proof is the finiteness of the number of ramified primes in the extension $L|K$ and the cohomological triviality of the unit group $U_{\mathfrak{P}}$ in an unramified local extension $L_{\mathfrak{P}}|K_{\mathfrak{p}}$. Nevertheless, Theorem (3.2) is fundamental for the idèle-theoretic development of global class field theory. In dimension $q = 0$ it allows us to prove the following corollary, which we refer to as the **Norm Theorem for Idèles:**

(3.4) Corollary. *An idèle $\mathfrak{a} \in I_K$ is the norm of an idèle $\mathfrak{b}$ of I_L if and only if each component $\mathfrak{a}_{\mathfrak{p}} \in K_{\mathfrak{p}}^{\times}$ is the norm of an element $\mathfrak{b}_{\mathfrak{P}} \in L_{\mathfrak{P}}^{\times}$ ($\mathfrak{P} \mid \mathfrak{p}$), i.e., if and only if it is a local norm everywhere.*

Proof. Note that $H^0(G, I_L) = I_L^G/N_G I_L = I_K/N_G I_L$ and $H^0(G_{\mathfrak{P}}, L_{\mathfrak{P}}^{\times}) = K_{\mathfrak{p}}^{\times}/N_{G_{\mathfrak{P}}} L_{\mathfrak{P}}^{\times}$. Thus by Theorem (3.2) we have $I_K/N_G I_L \cong \bigoplus_{\mathfrak{p}} K_{\mathfrak{p}}^{\times}/N_{G_{\mathfrak{P}}} L_{\mathfrak{P}}^{\times}$. If $\mathfrak{a} \in I_K$, then this isomorphism takes the 0-cohomology class $\mathfrak{a} \cdot N_G I_L = \overline{\mathfrak{a}}$ to its components $\overline{\mathfrak{a}}_{\mathfrak{p}}$, which by the Addendum (3.2) can be computed as $\overline{\mathfrak{a}}_{\mathfrak{p}} = \mathfrak{a}_{\mathfrak{p}} \cdot N_{G_{\mathfrak{P}}} L_{\mathfrak{P}}^{\times}$. Now, since we have an isomorphism, $\overline{\mathfrak{a}} = 1$ if and only if $\overline{\mathfrak{a}}_{\mathfrak{p}} = 1$, i.e., $\mathfrak{a} \in N_G I_L$ if and only if for every component $\mathfrak{a}_{\mathfrak{p}} \in N_{G_{\mathfrak{P}}} L_{\mathfrak{P}}^{\times}$.

The Norm Theorem for Idèles is an analogue of the "Hasse Norm Theorem", which states that if $L|K$ is a cyclic extension, an element $x \in K^{\times}$ is the norm of an element $y \in L^{\times}$ if and only if it is everywhere a local norm, i.e., is a norm for every extension $L_{\mathfrak{P}}|K_{\mathfrak{p}}$ (cf. §4, (4.8)). Contrary to the Norm Theorem for idèles, the Hasse Norm Theorem is a very deep number-theoretic result, for which up to date we do not have a direct proof. Corollary (3.4) only says that the element $x \in K^{\times}$, considered as a principal idèle, is the norm of an idèle $\mathfrak{b}$ of L; it leaves open the question whether this idèle can be chosen as a principal idèle $y \in L^{\times}$.

(3.5) Corollary. $H^1(G, I_L) = H^3(G, I_L) = 1$.

This follows from (3.2), since $H^1(G_{\mathfrak{P}}, L_{\mathfrak{P}}^{\times}) = H^1(G_{\mathfrak{P}}, L_{\mathfrak{P}}^{\times}) = 1$ for all $\mathfrak{P}$ (cf. II, (2.2) and II, (5.8)).

The fact that $H^1(G_{L|K}, I_L) = 1$ implies that the extensions $L|K$ form with respect to the idèle groups I_L a field formation in the sense of II, §1. This allows us to think of the cohomology groups $H^2(G_{L|K}, I_L)$ as the elements of

$$H^2(G_{\Omega|K}, I_{\Omega}) = \bigcup_L H^2(G_{L|K}, I_L), \quad {}^{9)}$$

---

9) Here Ω denotes the field of all algebraic numbers; however $H^2(G_{\Omega|K}, I_{\Omega})$ is used only as notation for the union on the right. If we consider I_{Ω} as the union of all I_L, or more precisely $I_{\Omega} = \varinjlim I_L$, then I_{Ω} is a $G_{\Omega|K}$-module, and we can define $H^2(G_{\Omega|K}, I_{\Omega})$ directly also for infinite Galois groups $G_{\Omega|K}$ (cf. [41]).

where the inclusions are given by the injective (because $H^1(G_{L|K}, I_L) = 1$) inflation maps. We will use this interpretation in all further considerations below. In particular, if $N \supseteq L \supseteq K$ are two normal extensions of K, then

$$H^2(G_{L|K}, I_L) \subseteq H^2(G_{N|K}, I_N) \subseteq H^2(G_{\Omega|K}, I_\Omega).$$

In local class field theory we have seen that the Brauer group $Br(K) = \bigcup_L H^2(G_{L|K}, L^\times)$ of a $\mathfrak{p}$-adic number field K is the union of the cohomology groups $H^2(G_{L|K}, L^\times)$ of the unramified extensions $L|K$, for which it is relatively easy to prove the reciprocity law. The role of the unramified extensions in the local theory is played in the global case by the cyclic **cyclotomic field extensions**, i.e., cyclic extensions which are contained in a field which is formed by adjoining roots of unity. We show already at this point the

(3.6) Theorem. *Let K be a finite algebraic number field. Then*

$$Br(K) = \bigcup_{L|K \text{ cyclic}} H^2(G_{L|K}, L^\times) \quad \text{and} \quad H^2(G_{\Omega|K}, I_\Omega) = \bigcup_{L|K \text{ cyclic}} H^2(G_{L|K}, I_L),$$

where $L|K$ ranges over all cyclic cyclotomic extensions.

For the proof we use the following

(3.7) Lemma. *Let K be a finite algebraic number field, S a finite set of primes of K, and m a natural number. Then there exists a cyclic cyclotomic field $L|K$ with the property that*

- $m \mid [L_{\mathfrak{P}} : K_{\mathfrak{p}}]$ *for all finite $\mathfrak{p} \in S$,*
- $[L_{\mathfrak{P}} : K_{\mathfrak{p}}] = 2$ *for all real-infinite $\mathfrak{p} \in S$.*

Proof. It suffices to prove the lemma for $K = \mathbb{Q}$, the general case follows from this by taking compositia. More precisely, if $N|\mathbb{Q}$ is a totally imaginary cyclic cyclotomic field such that for every prime number p above which there is a prime of K in S the degree $[N_{\mathfrak{P}} : \mathbb{Q}_p]$ is divisible by $m \cdot [K : \mathbb{Q}]$, then $L = K \cdot N$ has the desired property.

Let l^n be a prime power and let ζ be a primitive l^n-th root of unity. If $l \neq 2$, then the extension $\mathbb{Q}(\zeta)|\mathbb{Q}$ is cyclic of degree $l^{n-1} \cdot (l-1)$, and we denote the cyclic subfield of degree l^{n-1} by $L(l^n)$.

If $l = 2$, then the Galois group of $\mathbb{Q}(\zeta)|\mathbb{Q}$ is the direct product of a cyclic group of order 2 and a cyclic group of order 2^{n-2}. In this case we consider the field $L(2^n) = \mathbb{Q}(\xi)$ with $\xi = \zeta - \zeta^{-1}$. The automorphisms of $\mathbb{Q}(\zeta)$ are defined by $\sigma_\nu : \zeta \mapsto \zeta^\nu$, ν odd, and we have $\sigma_\nu(\xi) = \zeta^\nu - \zeta^{-\nu}$. Because $\zeta^{2^{n-1}} = -1$, $\sigma_\nu(\xi) = \sigma_{-\nu+2^{n-1}}(\xi)$, and since either ν or $-\nu + 2^{n-1} \equiv 1 \bmod (4)$, the automorphisms of $L(2^n) = \mathbb{Q}(\xi)$ are induced by those σ_ν with $\nu \equiv 1 \bmod (4)$.

Now an elementary calculation shows that the Galois group of $L(2^n)|\mathbb{Q}$ is cyclic of order 2^{n-2}. Moreover, because $\sigma_{-1}\xi = -\xi$, the field $L(l^n)$ is totally imaginary for large n.

If p is a prime number, then as n increases the local degree $[L(l^n)_{\mathfrak{P}} : \mathbb{Q}_p]$ becomes an arbitrarily large l-th power, since in any case $[\mathbb{Q}_p(\zeta) : \mathbb{Q}_p]$ becomes arbitrarily large, and we have $[\mathbb{Q}_p(\zeta) : L(l^n)_{\mathfrak{P}}] \leq l-1$, resp. ≤ 2 in case $l = 2$.

Now if $m = l_1^{r_1} \cdots l_s^{r_s}$, then the field

$$L = L(l_1^{n_1}) \cdots L(l_s^{n_s}) \cdot L(2^t)$$

has the desired properties, provided the n_i, t are chosen sufficiently large. In fact, for the finitely many prime numbers $p \in S$ the local degrees $[L_{\mathfrak{P}} : \mathbb{Q}_p]$ are divisible by every power $l_i^{r_i}$, and therefore by m; L is totally complex because of the factor $L(2^t)$, and L is cyclic over $\mathbb{Q}$, since the $L(l^n)$ are cyclic over $\mathbb{Q}$ with relatively prime degrees.

Proof of (3.6): We only give the proof for $H^2(G_{\Omega|K}, I_\Omega)$; the proof for $Br(K)$ is exactly the same if one replaces for the occurring fields L the idèle group I_L by the multiplicative group $L^\times$.

Let $c \in H^2(G_{\Omega|K}, I_\Omega)$, say $c \in H^2(G_{L'|K}, I_{L'})$, let m be the order of c and let S be the (finite) set of primes $\mathfrak{p}$ of K, for which the local components $c_\mathfrak{p}$ of c are not equal to 1. By the previous lemma there is a cyclic cyclotomic field $L|K$ with $m \mid [L_{\mathfrak{P}} : K_\mathfrak{p}]$ for the finite $\mathfrak{p} \in S$ and $[L_{\mathfrak{P}} : K_\mathfrak{p}] = 2$ for the real-infinite $\mathfrak{p} \in S$. If we form the compositum $N = L' \cdot L$, then we have

$$H^2(G_{L'|K}, I_{L'}) \quad \text{and} \quad H^2(G_{L|K}, I_L) \subseteq H^2(G_{N|K}, I_N),$$

and we will show that c lies in the group $H^2(G_{L|K}, I_L)$. Since the sequence

$$1 \longrightarrow H^2(G_{L|K}, I_L) \longrightarrow H^2(G_{N|K}, I_N) \xrightarrow{\text{res}_L} H^2(G_{N|L}, I_N)$$

is exact, it suffices to show that $\text{res}_L c = 1$. But by local class field theory, together with (3.2) and (3.3), we have $\text{res}_L c = 1 \Leftrightarrow (\text{res}_L c)_{\mathfrak{P}} = \text{res}_{L_{\mathfrak{P}}} c_\mathfrak{p} = 1$ for all primes $\mathfrak{P}$ of $L \Leftrightarrow \text{inv}_{N_{\mathfrak{P}'}|L_{\mathfrak{P}}}(\text{res}_{L_{\mathfrak{P}}} c_\mathfrak{p}) = [L_{\mathfrak{P}} : K_\mathfrak{p}] \cdot \text{inv}_{N_{\mathfrak{P}'}|K_\mathfrak{p}} c_\mathfrak{p} = \text{inv}_{N_{\mathfrak{P}'}|K_\mathfrak{p}} c_\mathfrak{p}^{[L_{\mathfrak{P}}:K_\mathfrak{p}]} = 0$ for all primes $\mathfrak{p}$ of $K \Leftrightarrow c_\mathfrak{p}^{[L_{\mathfrak{P}}:K_\mathfrak{p}]} = 1$ for all $\mathfrak{p} \in S$.

Now the last equality holds, because $c_\mathfrak{p}^m = 1$ and $m \mid [L_{\mathfrak{P}} : K_\mathfrak{p}]$ for the finite primes, and $[L_{\mathfrak{P}} : K_\mathfrak{p}] = 2$ for the real-infinite $\mathfrak{p} \in S$.

§ 4. Cohomology of the Idèle Class Group

The role of the multiplicative group of a field in the local theory is taken by the idèle class group in global class field theory. Thus our aim is to show that there is a canonical reciprocity isomorphism between the abelianization of the Galois group $G = G_{L|K}$ of a normal extension $L|K$ of finite algebraic number fields and the norm residue group $C_K/N_G C_L$; in other words: that the finite normal extensions $L|K$ of an algebraic number field K constitute a class formation in the sense of II, §1 with respect to the idèle class groups C_L.

In particular, we will have to prove that $H^1(G, C_L) = 1$ and that $H^2(G, C_L)$ is cyclic of order $[L : K]$. This will follow from the so-called first and second fundamental inequalities, which we prove below. The ideal-theoretic versions of these inequalities were already used in the proof of the main theorems of class field theory by TAKAGI (cf. [17], Teil I). Their proofs required, however, a considerable effort; in particular, the second inequality is proved using mainly Dirichlet series, thus analytic methods, which are redundant when working with idèles.

In what follows we fix a normal extension $L|K$ with a cyclic Galois group $G = G_{L|K}$ of prime order p. The **first fundamental inequality** is the relation

$$(C_K : N_G C_L) \geq p.$$

It follows immediately from the following

(4.1) Theorem. *The idèle class group C_L is a Herbrand module with Herbrand quotient*

$$h(C_L) = \frac{|H^0(G, C_L)|}{|H^1(G, C_L)|} = p \quad {}^{10)}.$$

From this we obtain as a

(4.2) Corollary.

$$|H^0(G, C_L)| = (C_K : N_G C_L) = |H^2(G, C_L)| = p \cdot |H^1(G, C_L)| \geq p.$$

Remark. If we knew that $H^1(G, C_L) = 1$, then Corollary (4.2) would immediately imply $|H^2(G, C_L)| = p$, thus that $H^2(G, C_L)$ is cyclic of order $[L : K]$. However, contrary to the cases where instead of C_L one considers the idèle group I_L (cf. (3.5)), or the multiplicative group $L^\times$ (Hilbert-Noether Theorem), this is not easy to show. That indeed $H^1(G, C_L) = 1$ will follow only from second fundamental inequality $(C_K : N_G C_L) = |H^0(G, C_L)| \leq p$ proved in the next section. Because of the isomorphism $H^1(G, C_L) \cong H^{-1}(G, C_L)$, it is easy to see that the statement $H^1(G, C_L) = 1$ is actually equivalent to Hasse's Norm Theorem (cf. (4.8)) mentioned in the previous section. This is the reason why a direct proof of Hasse's Norm Theorem would be very desirable; however, to date we do not have such a proof.

${}^{10)}$ See I, §6 for more on Herbrand quotients.

Proof of Theorem (4.1). Let S be a finite set of primes of K such that

1. S contains all infinite primes and all primes ramified in L,
2. $I_L = I_L^S \cdot L^\times$,
3. $I_K = I_K^S \cdot K^\times$.

Note that by (2.4) such a set S certainly exists. Then we have

$$C_L = I_L^S \cdot L^\times / L^\times \cong I_L^S / L^S,$$

where $L^S = L^\times \cap I_L^S$ is the group of S-units, i.e., the group of all those elements in $L^\times$ which are units for all the primes $\mathfrak{P}$ of L which do not lie above the primes in S (cf. §2, p. 118 and §1, p. 114). From I, (6.4) we obtain

$$h(C_L) = h(I_L^S) \cdot h(L^S)^{-1},$$

in the sense that when two of these Herbrand quotients are defined, then so is the third, and we have equality.

The proof splits into two parts, i.e., the computations of $h(I_L^S)$ and $h(L^S)$.

Because of Theorem (3.2), the computation of $h(I_L^S)$ is a local question. Let

n the number of primes in S,
N the number of primes of L, which lie above S, and
n_1 the number of primes in S, which are inert in L.

Since $[L : K]$ has prime degree, a prime of K that is not inert splits completely, i.e., decomposes into exactly p primes of L; thus $N = n_1 + p \cdot (n - n_1)$.

To compute the quotient $h(I_L^S) = |H^0(G, I_L^S)| / |H^1(G, I_L^S)|$, we have to determine $|H^0(G, I_L^S)|$ and $|H^1(G, I_L^S)|$. We do this making use of the isomorphism $H^q(G, I_L^S) \cong \prod_{\mathfrak{p} \in S} H^q(G_\mathfrak{P}, L_\mathfrak{P}^\times)$ from Theorem (3.2).

If $q = 1$, the above isomorphism immediately yields $H^1(G, I_L^S) = 1$, because $H^1(G_\mathfrak{P}, L_\mathfrak{P}^\times) = 1$. If $q = 0$, then $H^0(G, I_L^S) \cong \prod_{\mathfrak{p} \in S} H^0(G_\mathfrak{P}, L_\mathfrak{P}^\times)$, and it remains to determine the order of $H^0(G_\mathfrak{P}, L_\mathfrak{P}^\times)$, which is done using local class field theory. In fact, we have $H^0(G_\mathfrak{P}, L_\mathfrak{P}^\times) \cong G_\mathfrak{P}$ (cf. II, (5.9)), so that

$$|H^0(G_\mathfrak{P}, L_\mathfrak{P}^\times)| = \begin{cases} 1, & \text{if the prime } \mathfrak{p} \text{ lying under } \mathfrak{P} \text{ splits (because } G_\mathfrak{P} = 1), \\ p, & \text{if } \mathfrak{p} \text{ is inert (because } G_\mathfrak{P} = G). \end{cases}$$

Hence $|H^0(G, I_L^S)| = p^{n_1}$; since $H^1(G, I_L^S) = 1$, we have so $h(I_L^S) = p^{n_1}$.

For the computation of $h(L^S)$ we use the formula for the Herbrand quotient from Theorem I, (6.10). By (1.1), the group $L^S = L^\times \cap I_L^S$ of S-units of L is finitely generated of rank $N - 1$, and its fixed group $(L^S)^G = K^S = K^\times \cap L^S$ is the group of S-units of K and finitely generated of rank $n - 1$. Theorem I, (6.10) yields

$$h(L^S) = p^{(p(n-1)-N+1)/(p-1)} = p^{n_1-1}.$$

Since both Herbrand quotients $h(I_L^S)$ and $h(L^S)$ are defined, $h(C_L)$ is also defined, and the above formulas imply $h(C_L) = h(I_L^S) \cdot h(L^S)^{-1} = p$.

Theorem (4.1) has the following

(4.3) Corollary. *Let $L|K$ be a cyclic extension of prime power degree. Then K has infinitely many primes which are inert in L.*

Proof.   First let the degree $[L : K] = p$ be a prime number. We assume that the set $\mathfrak{U}$ of the primes of K remaining inert in L is finite. We show that under this assumption $C_K = N_G C_L$ $(G = G_{L|K})$, contradicting the first fundamental inequality (4.2). Let $\bar{a} \in C_K$ and let $\mathfrak{a} \in I_K$ be a representative idèle of $\bar{a}$ with local components $\mathfrak{a}_\mathfrak{p} \in K_\mathfrak{p}^\times$. The group of p-th powers $(K_\mathfrak{p}^\times)^p$ is open in $K_\mathfrak{p}^\times$ with respect to the valuation topology by II, (3.6). Therefore for each $\mathfrak{p} \in \mathfrak{U}$, $\mathfrak{a}_\mathfrak{p} \cdot (K_\mathfrak{p}^\times)^p$ is an open neighborhood of the element $\mathfrak{a}_\mathfrak{p}$, and since the field K is dense in its completion $K_\mathfrak{p}$, we can find an $x_\mathfrak{p} \in K^\times$ which lies in this neighborhood: $x_\mathfrak{p} \in \mathfrak{a}_\mathfrak{p} \cdot (K_\mathfrak{p}^\times)^p$. The Approximation Theorem in valuation theory implies further that there exists an $x \in K$, which approximates $x_\mathfrak{p}$ arbitrarily closely with respect to the prime $\mathfrak{p}$ for all $\mathfrak{p} \in \mathfrak{U}$. In particular, we may assume that along with $x_\mathfrak{p}$ we also have $x \in \mathfrak{a}_\mathfrak{p} \cdot (K_\mathfrak{p}^\times)^p$, and therefore $\mathfrak{a}_\mathfrak{p} \cdot x^{-1} \in (K_\mathfrak{p}^\times)^p$ for all $\mathfrak{p} \in \mathfrak{U}$. We now claim that the idèle $\mathfrak{a}' = \mathfrak{a} \cdot x^{-1}$ is the norm of an idèle $\mathfrak{b}$ in I_L. By (3.4) this is the case if and only if every component $\mathfrak{a}'_\mathfrak{p} \in K_\mathfrak{p}^\times$ is the norm of an element $\mathfrak{b}_\mathfrak{P} \in L_\mathfrak{P}^\times$ $(\mathfrak{P} \mid \mathfrak{p})$. For $\mathfrak{p} \in \mathfrak{U}$ this is true, because $[L_\mathfrak{P} : K_\mathfrak{p}] = p$ and $\mathfrak{a}'_\mathfrak{p} = \mathfrak{a}_\mathfrak{p} \cdot x^{-1} \in (K_\mathfrak{p}^\times)^p$, and for $\mathfrak{p} \notin \mathfrak{U}$ this is trivially the case, since $\mathfrak{p}$ splits completely because of the prime degree, and therefore $L_\mathfrak{P} = K_\mathfrak{p}$. Thus we have $\mathfrak{a}' = \mathfrak{a} \cdot x^{-1} = N_G \mathfrak{b}$, $\mathfrak{b} \in I_L$, from which we get $\bar{a} = \mathfrak{a} \cdot K^\times = \mathfrak{a}' \cdot K^\times = N_G \mathfrak{b} \cdot K^\times = N_G(\mathfrak{b} \cdot L^\times)$. Hence $C_K = N_G C_L$.

Now let $L|K$ be cyclic of degree p^r. We assume that almost all primes of K split in L. This means that the decomposition fields $Z_\mathfrak{P}$ are proper extensions of K for almost all primes $\mathfrak{P}$ of L, and therefore in each case contain a field L_0 between K and L of degree p. But in the cyclic extension $L|K$ there is only one field L_0 of degree p, which is therefore contained in almost all decomposition fields $Z_\mathfrak{P}$. This implies that almost all primes $\mathfrak{p}$ of K decompose in the cyclic extension L_0 of degree p, which contradicts the first part of the proof.

We now prove the **second fundamental inequality** $(C_K : N_G C_L) \leq p$ for cyclic extensions $L|K$ of prime degree, making the additional assumption that K contains the p-th roots of unity. In this case L is a Kummer extension: $L = K(\sqrt[p]{x_0})$, $x_0 \in K^\times$. We start with the following lemma:

(4.4) Lemma. *Let $N = K(\sqrt[p]{x})$, $x \in K^\times$, be any Kummer extension over K, and let $\mathfrak{p}$ be a finite prime of K not lying over the prime number p. Then $\mathfrak{p}$ is unramified in N if and only if $x \in U_\mathfrak{p} \cdot (K_\mathfrak{p}^\times)^p$, and $\mathfrak{p}$ splits completely in N if and only if $x \in (K_\mathfrak{p}^\times)^p$.*

Proof.   Let $\mathfrak{P}$ be a prime of N over $\mathfrak{p}$. Then $N_\mathfrak{P} = K_\mathfrak{p}(\sqrt[p]{x})$. If $x = u \cdot y^p$, $u \in U_\mathfrak{p}$, $y \in K_\mathfrak{p}^\times$, then $N_\mathfrak{P} = K_\mathfrak{p}(\sqrt[p]{x}) = K_\mathfrak{p}(\sqrt[p]{u})$. If the equation $X^p - u = 0$

is irreducible over the residue field of $K_\mathfrak{p}$, then it is also irreducible over $K_\mathfrak{p}$, and $N_\mathfrak{P}|K_\mathfrak{p}$ is an unramified extension of degree p. If $X^p - u = 0$ is reducible over the residue field of $K_\mathfrak{p}$, then it splits into p distinct linear factors there, since p is distinct from the characteristic of the residue field, and by Hensel's Lemma $X^p - u = 0$ also splits into linear factors over $K_\mathfrak{p}$, so that $N_\mathfrak{P} = K_\mathfrak{p}$. In both cases $N_\mathfrak{P}|K_\mathfrak{p}$ is unramified, i.e., $\mathfrak{p}$ is unramified in N.

Conversely, if $\mathfrak{p}$ is unramified in N, then $N_\mathfrak{P} = K_\mathfrak{p}(\sqrt[p]{x})$ is unramified over $K_\mathfrak{p}$, and we have $\sqrt[p]{x} = u \cdot \pi^k$, where $u \in U_\mathfrak{P}$ and $\pi \in K_\mathfrak{p}$ is a prime element (of smallest value 1). Thus we have $x = u^p \cdot \pi^{k \cdot p}$, and therefore $u^p \in U_\mathfrak{p}$, $\pi^{k \cdot p} \in (K_\mathfrak{p}^\times)^p$, i.e., $x \in U_\mathfrak{p} \cdot (K_\mathfrak{p}^\times)^p$.

The prime $\mathfrak{p}$ decomposes in N if and only if $N_\mathfrak{P} = K_\mathfrak{p}(\sqrt[p]{x}) = K_\mathfrak{p}$, hence if and only if $x \in (K_\mathfrak{p}^\times)^p$.

(4.5) Theorem. *Let $L|K$ be a cyclic extension of prime degree p. Assume the field K contains the p-th roots of unity. Then*

$$|H^0(G, C_L)| = (C_K : N_G C_L) \leq p.$$

The difficulty here is that that we cannot a priori decide which idèle classes in C_K are represented by a norm idèle, and therefore lie in $N_G C_L$. This is completely different from the case of idèle groups, where by the Norm Theorem for idèle groups $\mathfrak{a} \in I_K$ is a norm if and only if it is a local norm everywhere (cf. (3.4)). We work around this by considering instead of $N_G C_L$ an auxiliary group $\bar{F}$ which is constructed such that its elements are represented by norm idèles, hence $\bar{F} \subseteq N_G C_L$, and which has the property that its index $(C_K : \bar{F})$ can actually be shown to be equal to p. Using this $\bar{F}$, we obtain the inequality

$$(C_K : N_G C_L) \leq (C_K : \bar{F}) = p.$$

Let $L = K(\sqrt[p]{x_0})$, $x_0 \in K^\times$. Let S be a finite set of primes of K such that

1. S contains all the primes above p and all infinite primes of K,
2. $I_K = I_K^S \cdot K^\times$,
3. $x_0 \in K^S = I_K^S \cap K^\times$ (i.e., x_0 is an S-unit).

Here 2. can be satisfied by (2.4), and 3. because x_0 is a unit for almost all primes.

Together with S we choose m additional primes $\mathfrak{q}_1, \ldots, \mathfrak{q}_m \notin S$ that split completely in L; set $S^* = S \cup \{\mathfrak{q}_1, \ldots, \mathfrak{q}_m\}$. To construct $\bar{F}$, we have to specify an idèle group $F \subseteq I_K$ whose elements represent the idèle classes of $\bar{F}$. It must consist of nothing but norm idèles so that $\bar{F} \subseteq N_G C_L$, it must be sufficiently large to ensure that the index $(C_K : \bar{F})$ is finite, and it must be simple enough so that it is possible to compute this index. These properties are satisfied by the idèle group

$$F = \prod_{\mathfrak{p} \in S} (K_\mathfrak{p}^\times)^p \times \prod_{i=1}^m K_{\mathfrak{q}_i}^\times \times \prod_{\mathfrak{p} \notin S^*} U_\mathfrak{p} \qquad {}^{11)}.$$

To see that $F \subseteq N_G I_L$, it suffices by the Norm Theorem for idèles to convince ourselves that the components $\mathfrak{a}_\mathfrak{p}$ of each idèle $\mathfrak{a} \in F$ are norms from the extension $L_\mathfrak{P} | K_\mathfrak{p}$ ($\mathfrak{P} | \mathfrak{p}$).

This is true for $\mathfrak{p} \in S$, because $\mathfrak{a}_\mathfrak{p} \in (K_\mathfrak{p}^\times)^p \subseteq N_{L_\mathfrak{P}|K_\mathfrak{p}} L_\mathfrak{P}^\times$ (regardless of $[L_\mathfrak{P} : K_\mathfrak{p}] = p$ or $= 1$); this is trivially true for $\mathfrak{p} = \mathfrak{q}_i$, because $\mathfrak{q}_i$ splits completely so that $L_\mathfrak{P} = K_\mathfrak{p}$; and it is true for $\mathfrak{p} \notin S^*$, because $x_0 \in U_\mathfrak{p}$ by 3. and therefore by Lemma (4.4) each $\mathfrak{p} \notin S^*$ is unramified in $L = K(\sqrt[p]{x_0})$, so that $\mathfrak{a}_\mathfrak{p} \in U_\mathfrak{p} \subseteq N_{L_\mathfrak{P}|K_\mathfrak{p}} L_\mathfrak{P}^\times$ by II, (4.4). If we now set $\bar{F} = F \cdot K^\times / K^\times$, then $\bar{F} \subseteq N_G C_L$, since each idèle class $\bar{\mathfrak{a}}$ is represented by a norm idèle $\mathfrak{a} \in F$. To compute the index $(C_K : \bar{F})$, we consider the following decomposition:

$$(C_K : \bar{F}) = (I_K^{S^*} \cdot K^\times / K^\times : F \cdot K^\times / K^\times) = (I_K^{S^*} \cdot K^\times : F \cdot K^\times) =$$
$$(I_K^{S^*} : F) / ((I_K^{S^*} \cap K^\times) : (F \cap K^\times)) \qquad {}^{12)}.$$

It allows us to split the computation of $(C_K : \bar{F})$ into two parts, the computation of $(I_K^{S^*} : F)$, which is of a purely local nature, and the computation of $((I_K^{S^*} \cap K^\times) : (F \cap K^\times))$, which uses global considerations.

I. We have $(I_K^{S^*} : F) = \prod_{\mathfrak{p} \in S} (K_\mathfrak{p}^\times : (K_\mathfrak{p}^\times)^p)$; since $S \subseteq S^\times$, the map

$$I_K^{S^*} \longrightarrow \prod_{\mathfrak{p} \in S} K_\mathfrak{p}^\times / (K_\mathfrak{p}^\times)^p \quad \text{with} \quad \mathfrak{a} \longmapsto \prod_{\mathfrak{p} \in S} \mathfrak{a}_\mathfrak{p} \cdot (K_\mathfrak{p}^\times)^p$$

is trivially surjective, and its kernel consists precisely of those idèles $\mathfrak{a} \in I_K^{S^*}$ for which $\mathfrak{a}_\mathfrak{p} \in (K_\mathfrak{p}^\times)^p$ for $\mathfrak{p} \in S$ lie in the kernel; i.e., the idèles in F. By the local theory (cf. II, (3.7)) we have

$$(K_\mathfrak{p}^\times : (K_\mathfrak{p}^\times)^p) = p^2 \cdot |p|_\mathfrak{p}^{-1},$$

so that $(I_K^{S^*} : F) = p^{2n} \cdot \prod_{\mathfrak{p} \in S} |p|_\mathfrak{p}^{-1}$, where n is the number of primes in S. Since the primes $\mathfrak{p} \notin S$ do not lie above the prime number p, $|p|_\mathfrak{p} = 1$ for $\mathfrak{p} \notin S$, and by the product formula $\prod_{\mathfrak{p} \in S} |p|_\mathfrak{p} = \prod_\mathfrak{p} |p|_\mathfrak{p} = 1$, hence $(I_K^{S^*} : F) = p^{2n}$.

---

[11] That we consider precisely this idèle group is motivated as follows: If we start heuristically with the reciprocity law, which has not been proved yet, we see that the Kummer extension $T = K(\sqrt[p]{K^S})$ has as norm group the idèle class group $\bar{E} = E \cdot K^\times / K^\times$ formed from $E = \prod_{\mathfrak{p} \in S} (K_\mathfrak{p}^\times)^p \times \prod_{\mathfrak{p} \notin S} U_\mathfrak{p}$ (cf. (7.7)). By inserting additional factors $K_{\mathfrak{q}_i}^\times$, i.e., choosing suitable primes $\mathfrak{q}_i$, we try to enlarge $\bar{E}$ to a group $\bar{F}$ such that $\bar{F}$ becomes the norm group of the field $L = K(\sqrt[p]{x_0}) \subseteq T$.

[12] The last of these equations results from a general elementary group-theoretical fact: If $B \subseteq A$, C are subgroups of an abelian group, then the canonical surjective homomorphism $A/B \to A \cdot C / B \cdot C$ has kernel $A \cap B \cdot C / B \cong A \cap C / B \cap C$.

II. An elementary calculation shows that

$$((I_K^{S^*} \cap K^\times) : (F \cap K^\times)) = (K^{S^*} : (F \cap K^\times))$$
$$= (K^{S^*} : (K^{S^*})^p)/((F \cap K^\times) : (K^{S^*})^p),$$

where K^{S^*} is the group of S^*-units. By (1.1) this group is finitely generated of rank $n+m-1$ ($n+m$ is the number of primes in S^*). Moreover, K^{S^*} contains the p-th roots of unity, and it easily follows that $(K^{S^*} : (K^{S^*})^p) = p^{n+m}$.

Altogether we therefore have

$$(C_K : N_G C_L) \le (C_K : \bar{F}) = p^{n-m} \cdot ((F \cap K^\times) : (K^{S^*})^p),$$

and the second fundamental inequality is proved, provided that we can choose the primes $\mathfrak{q}_1, \ldots, \mathfrak{q}_m$ splitting in L in such a way that $m = n - 1$, and

$$K^\times \cap F = K^\times \cap \Big(\prod_{\mathfrak{p} \in S} (K_\mathfrak{p}^\times)^p \times \prod_{i=1}^m K_{\mathfrak{q}_i}^\times \times \prod_{\mathfrak{p} \notin S^*} U_\mathfrak{p} \Big) =$$
$$= K^\times \cap \bigcap_{\mathfrak{p} \in S} (K_\mathfrak{p}^\times)^p \cap \bigcap_{i=1}^m K_{\mathfrak{q}_i}^\times \cap \bigcap_{\mathfrak{p} \notin S^*} U_\mathfrak{p} =$$
$$= K^\times \cap \bigcap_{\mathfrak{p} \in S} (K_\mathfrak{p}^\times)^p \cap \bigcap_{\mathfrak{p} \notin S^*} U_\mathfrak{p} = (K^{S^*})^p;$$

using Lemma (4.4), we formulate this as follows:

Sublemma. *There exist* $n - 1$ *primes of* K, $\mathfrak{q}_1, \ldots, \mathfrak{q}_{n-1} \notin S$ *that split completely in* L *and satisfy the following condition:*

If $N = K(\sqrt[p]{x})$ *is a Kummer extension over* K *in which all* $\mathfrak{p} \in S$ *split completely and all* $\mathfrak{p} \neq \mathfrak{q}_1, \ldots, \mathfrak{q}_{n-1}$ *are unramified, then* $N = K(\sqrt[p]{x}) = K$.

In fact, the desired equality

$$K^\times \cap \bigcap_{\mathfrak{p} \in S} (K_\mathfrak{p}^\times)^p \cap \bigcap_{\mathfrak{p} \notin S^*} U_\mathfrak{p} = (K^{S^*})^p$$

follows immediately from this. The inclusion $\supseteq$ is trivial. Let $x \in K^\times \cap \bigcap_{\mathfrak{p} \in S} (K_\mathfrak{p}^\times)^p \cap \bigcap_{\mathfrak{p} \notin S^*} U_\mathfrak{p}$ and $N = K(\sqrt[p]{x})$. By (4.4) every $\mathfrak{p} \in S$ splits completely in N, since $x \in (K_\mathfrak{p}^\times)^p$. For $\mathfrak{p} \notin S^*$ we have $x \in U_\mathfrak{p} \subseteq U_\mathfrak{p} \cdot (K_\mathfrak{p}^\times)^p$, so that every $\mathfrak{p} \notin S^*$ is unramified in N by (4.4). Hence the sublemma yields $N = K(\sqrt[p]{x}) = K$, so that $x \in (K^\times)^p$, and because $x \in U_\mathfrak{p}$ for $\mathfrak{p} \notin S^*$, x lies in $(K^\times)^p \cap K^{S^*} = (K^{S^*})^p$.

The statement about the prime decomposition in Kummer extensions given by the sublemma represents the global part of the proof of the second fundamental inequality. To prove it, we consider the field $T = K(\sqrt[p]{K^S})$ obtained by adjoining the p-th roots of all the elements in the group K^S. By (1.3)

$$\chi(G_{T|K}) \cong K^S \cdot (K^\times)^p/(K^\times)^p \cong K^S/(K^S)^p.$$

Since K^S is finitely generated of rank $n - 1 = |S| - 1$ (cf. (1.1)) and contains the p-th roots of unity, we have $K^S \cong (\zeta) \times \mathbb{Z} \times \cdots \times \mathbb{Z}$; also $K^S/(K^S)^p \cong (\zeta) \times \mathbb{Z}/p\mathbb{Z} \times \cdots \times \mathbb{Z}/p\mathbb{Z}$ (ζ a primitive p-th root of unity). Thus the Galois group $G_{T|K}$ is the direct product of cyclic groups $\mathfrak{Z}_i$ of order p, $G_{T|K} = \mathfrak{Z}_1 \times \cdots \times \mathfrak{Z}_n$, and we have for the field degree $[T : K] = (K^S : (K^S)^p) = p^n$.

The field $L = K(\sqrt[p]{x_0})$ lies in T because $x_0 \in K^S$, and it is easy to see that we can assume $G_{T|L} = \mathfrak{Z}_1 \times \cdots \times \mathfrak{Z}_{n-1}$. Let $T_i \subseteq T$ be the fixed field of $\mathfrak{Z}_i \subseteq G_{T|K}$, thus $G_{T|T_i} = \mathfrak{Z}_i$.

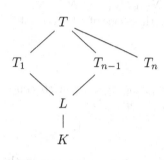

Because $\mathfrak{Z}_i \subseteq G_{T|L}$ we have $L \subseteq T_i$ for $i = 1, \ldots, n-1$, and the given fields form an arrangement as in the diagram on the left. Now we choose for each $i = 1, \ldots, n$ a prime $\mathfrak{Q}_i$ of the field T_i, such that the $\mathfrak{Q}_i$ are inert in T, and such that the primes $\mathfrak{q}_1, \ldots, \mathfrak{q}_n$ lying under the $\mathfrak{Q}_i$ are all distinct and are not in S. This is possible by (4.3). We claim that the $\mathfrak{q}_1, \ldots, \mathfrak{q}_{n-1} \notin S$ satisfy the conditions of the lemma.

To prove that $\mathfrak{q}_1, \ldots, \mathfrak{q}_{n-1}$ split completely in L, we observe that T_i is the decomposition field of the unique extension $\mathfrak{Q}'_i$ of $\mathfrak{Q}_i$ to T over K, $i = 1, \ldots, n$. This decomposition field Z_i is contained in T_i, since $\mathfrak{Q}_i$ is inert in T. On the other hand, by (4.4), $\mathfrak{q}_i$ is unramified in every field $K(\sqrt[p]{x})$, $x \in K^S$, thus also in T, so that the Galois group $G_{T|Z_i}$ of $T|Z_i$ is isomorphic to the Galois group of the residue field extension of $T|Z_i$, and therefore is cyclic. But a generator of $G_{T|Z_i}$ has order p as an element of $G_{T|K}$, hence $[T : Z_i] = p$ and $Z_i = T_i$. Since L is contained in the decomposition fields T_i for $i = 1, \ldots, n-1$, it follows that the primes $\mathfrak{q}_1, \ldots, \mathfrak{q}_{n-1}$ split completely in L.

Set $U_i = U_{\mathfrak{q}_i}$, $i = 1, \ldots, n$. We show next that the homomorphism

$$K^S/(K^S)^p \longrightarrow \prod_{i=1}^{n} U_i/(U_i)^p, \quad x \cdot (K^S)^p \mapsto \prod_{i=1}^{n} x \cdot (U_i)^p \quad (x \in K^S),$$

is bijective. For injectivity note that if $x \in (U_i)^p \subseteq (K_{\mathfrak{q}_i}^\times)^p$, then it follows from (4.4) that the primes $\mathfrak{q}_1, \ldots, \mathfrak{q}_n$ split completely in the field $K(\sqrt[p]{x})$, so that $K(\sqrt[p]{x})$ is contained in the decomposition fields T_i, $i = 1, \ldots, n$. Hence $K(\sqrt[p]{x}) \subseteq \bigcap_{i=1}^{n} T_i = K$, and $x \in (K^\times)^p \cap K^S = (K^S)^p$.

We show surjectivity by comparing orders. We know that $(K^S : (K^S)^p) = p^n$; on the other hand, using II, (3.8) we find $(U_i : (U_i)^p) = p \cdot |p|_{\mathfrak{q}_i}^{-1} = p$, i.e., $(K^S : (K^S)^p)$ and $\prod_{i=1}^{n} U_i/(U_i)^p$ have the same order p^n.

Now let $N = K(\sqrt[p]{x})$, $x \in K^\times$ be a Kummer extension in which the $\mathfrak{p} \in S$ split completely and the $\mathfrak{p} \neq \mathfrak{q}_1, \ldots, \mathfrak{q}_{n-1}$ are unramified. To prove that $N = K$, it suffices by (4.2) to show that $C_K = N_{N|K} C_N$. Let $\bar{\mathfrak{a}} \in C_K = I_K/K^\times = I_K^S \cdot K^\times/K^\times$, and let $\mathfrak{a} \in I_K^S$ be a representative of the class $\bar{\mathfrak{a}}$. If we set

$\bar{a}_i = a_{q_i} \cdot (U_i)^p$ $(a_{q_i} \in U_i)$, $i = 1, \ldots, n$, then the fact that $K^S/(K^S)^p \longrightarrow$ $\prod_{i=1}^{n} U_i/(U_i)^p$ is surjective implies that there is a $y \in K^S$ with $y \cdot (U_i)^p = \bar{a}_i$, so that $a_{q_i} = y \cdot u_i^p$, $u_i \in U_i$ for $i = 1, \ldots, n$. The idèle $a' = a \cdot y^{-1}$ belongs to the same idèle class as a, and it follows easily from from the Norm Theorem for idèles that moreover $a' \in N_{N|K}I_N$: If $p \in S$, then a'_p is a norm, since p splits completely in N, and therefore $N_{\mathfrak{P}} = K_p$ ($\mathfrak{P} \mid p$). For the primes q_i, $i = 1, \ldots, n-1$, $a'_{q_i} = u_i^p$ as the p-th power of a norm, and for $p \notin S$, $p \neq q_1, \ldots, q_{n-1}$, the norm property of a'_p follows by II, (4.4) from the fact that $a'_p \in U_p$ is a unit and by assumption p is in N, i.e., $N_{\mathfrak{P}}|K_p$ ($\mathfrak{P} \mid p$) is unramified. Therefore we obtain $\bar{a} = a' \cdot K^\times \in N_{N|K}C_N$, i.e., $N = K$. This proves the sublemma and consequently completes the proof of Theorem (4.5).

Theorems (4.1) and (4.5) together imply

(4.6) Corollary. *If $L|K$ is a cyclic extension of prime degree p with Galois group $G = G_{L|K}$, and K contains the p-th roots of unity, then*

$$H^0(G, C_L) \cong H^2(G, C_L) \cong G \quad \text{and} \quad H^1(G, C_L) = 1.$$

From what we have shown for Kummer extensions, it is easy to prove the following, more general result.

(4.7) Theorem. *If $L|K$ is a normal extension with Galois group $G = G_{L|K}$, then we have $H^1(G, C_L) = 1$.*

We prove this by induction on the order n of the group of G. The case $n = 1$ is trivial. Let us assume that $H^1(G, C_L) = 1$ for every extension $L|K$ of degree $< n$. If the order $n = |G|$ is not a p-power, then each p-Sylow subgroup G_p of G has order smaller than n, so that by the induction hypothesis $H^1(G_p, C_L) = 1$, and therefore $H^1(G, C_L) = 1$ by I, (4.17).

Thus is suffices to prove this for a p-group G. In this case, let $g \subseteq G$ be a normal subgroup of index p; g is the Galois group of an intermediate field M, $K \subseteq M \subseteq L$, $g = G_{L|M}$. Now if $p < n$, then by assumption $H^1(G/g, C_M) = H^1(g, C_L) = 1$, and from the exact sequence (cf. I, (4.6))

$$1 \longrightarrow H^1(G/g, C_M) \xrightarrow{\text{inf}} H^1(G, C_L) \xrightarrow{\text{res}} H^1(g, C_L)$$

we see that $H^1(G, C_L) = 1$.

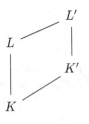

Assume $p = n$. In order to be able to apply Corollary (4.6), we replace K by the extension K' obtained by adjoining a primitive p-th root of unity to K, and set $L' = L \cdot K'$. Obviously $[K' : K] \leq p - 1 < p = n$ and $[L' : K'] = p$. Because $[K' : K] < n$, resp. (4.6), we have $H^1(G_{K'|K}, C_{K'}) = H^1(G_{L'|K'}, C_{L'}) = 1$, and from the exact sequence

$$1 \longrightarrow H^1(G_{K'|K}, C_{K'}) \xrightarrow{\text{inf}} H^1(G_{L'|K}, C_{L'}) \xrightarrow{\text{res}} H^1(G_{L'|K'}, C_{L'})$$

we obtain $H^1(G_{L'|K}, C_{L'}) = 1$. On the other hand, because the sequence

$$1 \longrightarrow H^1(G, C_L) \xrightarrow{\text{inf}} H^1(G_{L'|K}, C_{L'}) = 1$$

is also exact, we see that $H^1(G_{L'|K}, C_{L'}) = 1$ implies $H^1(G, C_L) = 1$.

For cyclic extensions Theorem (4.7) is just another form of the **Hasse Norm Theorem** mentioned earlier:

(4.8) Corollary. *If the extension $L|K$ is cyclic, then an element $x \in K^\times$ is a norm if and only if it is locally a norm everywhere.*

Proof. The sequence of G-modules $1 \to L^\times \to I_L \to C_L \to 1$ yields the exact cohomology sequence

$$H^{-1}(G, C_L) \longrightarrow H^0(G, L^\times) \longrightarrow H^0(G, I_L) \cong \bigoplus_{\mathfrak{p}} H^0(G_{\mathfrak{P}}, L_{\mathfrak{P}}^\times).$$

Since G is cyclic, $H^{-1}(G, C_L) \cong H^1(G, C_L) = 1$ by Theorem (4.7), which implies that the canonical homomorphism

$$K^\times / N_{L|K} L^\times \longrightarrow \bigoplus_{\mathfrak{p}} K_{\mathfrak{p}}^\times / N_{L_{\mathfrak{P}}|K_{\mathfrak{p}}} L_{\mathfrak{P}}^\times$$

is injective; this is precisely the assertion of the Hasse Norm Theorem.

(4.9) Theorem. *Let $L|K$ be a normal extension with Galois group $G = G_{L|K}$. Then the order of $H^2(G, C_L)$ is a divisor of the degree $[L : K]$.*

We prove this again by induction on the order n of the group G. For $n = 1$ the theorem is trivial, and we assume it holds for all normal extensions of degree less than n. If the order of $|G| = n$ is not a prime power, every p-Sylow subgroup G_p of G has a smaller order that n, and by assumption the order $|H^2(G_p, C_L)|$ divides the order n_p of G_p, i.e., the maximal p-power dividing n. If $H_p^2(G, C_L)$ denotes the p-Sylow subgroup of $H^2(G, C_L)$ [13], then the restriction map

$$H_p^2(G, C_L) \xrightarrow{\text{res}} H^2(G_p, C_L)$$

is injective by I, (4.16). Hence the order $|H_p^2(G, C_L)|$ divides the maximal p-power n_p which divides n, and since $H^2(G, C_L)$ is a direct product of its p-Sylow groups, $|H^2(G, C_L)|$ is a divisor of n.

---

[13] $H_p^2(G, C_L)$ consists precisely of those elements of $H^2(G, C_L)$ which have p-power order; $H_p^2(G, C_L)$ is often referred to as the p-primary component of $H^2(G, C_L)$.

Thus we can again assume G is a p-group. Choose a normal subgroup $g \subseteq G$ of index p. Because $|g| = n/p < n$, the order of $H^2(g, C_L)$ is a divisor of n/p, and taking into account that $H^1(g, C_L) = 1$, I, (4.7) shows that the sequence

$$1 \longrightarrow H^2(G/g, C_L^g) \xrightarrow{\text{inf}} H^2(G, C_L) \xrightarrow{\text{res}} H^2(g, C_L)$$

is exact. But G/g is the Galois group of a cyclic extension of prime degree $L'|K$, $K \subseteq L' \subseteq L$, i.e., $G/g = G_{L'|K}$, $C_L^g = C_{L'}$, and by (4.6) we have $|H^2(G/g, C_L^g)| = p$. It follows from the above exact sequence that $|H^2(G, C_L)|/p$ divides n/p, and therefore $|H^2(G, C_L)|$ divides n, as claimed.

With Theorem (4.9) we have not yet reached our goal to show that $H^2(G, C_L)$ is cyclic of the same order as $[L : K]$. To show this, we will associate with the group $H^2(G, C_L)$ an invariant homomorphism, as required by Axiom II for class formations.

§ 5. Idèle Invariants

Recall that we want to show that the extensions $L|K$ form a class formation in the sense of II, §1 with respect to the idèle class group C_L. With Theorem (4.7) we have shown that Axiom I is satisfied. Hence it remains to show that for every normal extension $L|K$ there is an invariant isomorphism

$$H^2(G_{L|K}, C_L) \longrightarrow \tfrac{1}{[L:K]}\mathbb{Z}/\mathbb{Z}$$

which satisfies the compatibility properties required by Axiom II. It is, of course, essential that we construct the invariant isomorphism in a canonical way to also obtain a canonical reciprocity law, the **Artin reciprocity law**. In a certain sense we will retrieve this invariant map, and with it the reciprocity law, from the local theory, by relating the group $H^2(G_{L|K}, C_L)$ to the group $H^2(G_{L|K}, I_L)$ formed with the idèle group I_L as the underlying module. For the latter group we obtain from the decomposition $H^2(G_{L|K}, I_L) \cong \bigoplus_{\mathfrak{p}} H^2(G_{L_{\mathfrak{P}}|K_{\mathfrak{p}}}, L_{\mathfrak{P}}^\times)$ immediately an invariant map from local class field theory by taking the sum of the canonical invariant isomorphisms of the local extensions $L_{\mathfrak{P}}|K_{\mathfrak{p}}$. We will show that the invariants of the elements in $H^2(G_{L|K}, I_L)$ yield invariants for the elements of $H^2(G_{L|K}, C_L)$.

Let $L|K$ be a normal extension of finite algebraic number fields, and let $G_{L|K}$ be its Galois group. By (3.2) we have the decomposition

$$H^2(G_{L|K}, I_L) \cong \bigoplus_{\mathfrak{p}} H^2(G_{L_{\mathfrak{P}}|K_{\mathfrak{p}}}, L_{\mathfrak{P}}^\times),$$

where $\bigoplus$ again denotes the direct sum. For every prime $\mathfrak{p}$ of K we have from local class field theory the isomorphism (cf. II, (5.5))

$$\mathrm{inv}_{L_{\mathfrak{P}}|K_{\mathfrak{p}}} : H^2(G_{L_{\mathfrak{P}}|K_{\mathfrak{p}}}, L_{\mathfrak{P}}^{\times}) \longrightarrow \tfrac{1}{[L_{\mathfrak{P}}:K_{\mathfrak{p}}]}\mathbb{Z}/\mathbb{Z} \subseteq \tfrac{1}{[L:K]}\mathbb{Z}/\mathbb{Z} \qquad (\mathfrak{P} \mid \mathfrak{p}) \ ^{14)}$$

The local invariant isomorphism $\mathrm{inv}_{L_{\mathfrak{P}}|K_{\mathfrak{p}}}$ is the composition of three homomorphisms; however, for now it is not necessary for us to know this map explicitly, at this point it is only important that we know that it satisfies the compatibility conditions from Axiom II of class formations (cf. II, (5.6)).

(5.1) Definition. *If* $c_{\mathfrak{p}} \in H^2(G_{L_{\mathfrak{P}}|K_{\mathfrak{p}}}, L_{\mathfrak{P}}^{\times})$ ($\mathfrak{P}$ *an arbitrarily chosen prime over* $\mathfrak{p}$*) are the local components of* $c \in H^2(G_{L|K}, I_L)$*, then we set*

$$\mathrm{inv}_{L|K}c = \sum_{\mathfrak{p}} \mathrm{inv}_{L_{\mathfrak{P}}|K_{\mathfrak{p}}} c_{\mathfrak{p}} \in \tfrac{1}{[L:K]}\mathbb{Z}/\mathbb{Z}.$$

Note that here almost all $c_{\mathfrak{p}} = 1$, so that the sum contains only finitely many non-zero summands. In particular, we obtain an invariant homomorphism

$$\mathrm{inv}_{L|K} : H^2(G_{L|K}, I_L) \longrightarrow \tfrac{1}{[L:K]}\mathbb{Z}/\mathbb{Z}.$$

(5.2) Proposition. *If* $N \supseteq L \supseteq K$ *are normal extensions of the field* K*, then*

$$\begin{aligned}
\mathrm{inv}_{N|K}c &= \mathrm{inv}_{L|K}c, & c \in H^2(G_{L|K}, I_L) \subseteq H^2(G_{N|K}, I_N), \\
\mathrm{inv}_{N|L}(\mathrm{res}_L c) &= [L:K] \cdot \mathrm{inv}_{N|K}c, & c \in H^2(G_{N|K}, I_N), \\
\mathrm{inv}_{N|K}(\mathrm{cor}_K c) &= \mathrm{inv}_{N|L}c, & c \in H^2(G_{N|L}, I_N).
\end{aligned}$$

The last two formulas require only that $N|K$ *be normal.*

Here we use the convention made after (3.5) to interpret the inflation map

$$H^2(G_{L|K}, I_L) \longrightarrow H^2(G_{N|K}, I_N) \qquad (N \supseteq L \supseteq K)$$

as an inclusion, so that $H^2(G_{L|K}, I_L) \subseteq H^2(G_{N|K}, I_N)$.

Proof.   The proposition follows from the behavior of the local invariants with respect to the maps incl, res and cor. If $c \in H^2(G_{L|K}, I_L)$, then by (3.3)

$$\mathrm{inv}_{N|K}c = \sum_{\mathfrak{p}} \mathrm{inv}_{N_{\mathfrak{P}'}|K_{\mathfrak{p}}} c_{\mathfrak{p}} = \sum_{\mathfrak{p}} \mathrm{inv}_{L_{\mathfrak{P}}|K_{\mathfrak{p}}} c_{\mathfrak{p}} = \mathrm{inv}_{L|K}c$$

Here $\mathfrak{P}'$ is an arbitrary prime of N over $\mathfrak{p}$ and $\mathfrak{P}$ is the prime of L lying under $\mathfrak{P}'$. If $c \in H^2(G_{N|K}, I_N)$, and $\mathfrak{P}$ runs through the primes of L, then

---

14) This invariant isomorphism is independent of the choice of $\mathfrak{P} \mid \mathfrak{p}$ in the following sense: If $\mathfrak{P}' \mid \mathfrak{p}$ is another prime of L, then the canonical $K_{\mathfrak{p}}$-isomorphism $L_{\mathfrak{P}} \rightarrow L_{\mathfrak{P}'}$ yields a canonical isomorphism between $H^2(G_{L_{\mathfrak{P}}|K_{\mathfrak{p}}}, L_{\mathfrak{P}}^{\times})$ and $H^2(G_{L_{\mathfrak{P}'}|K_{\mathfrak{p}}}, L_{\mathfrak{P}'}^{\times})$, which trivially preserves the invariant map.

$$\mathrm{inv}_{N|L}(\mathrm{res}_L c) = \sum_{\mathfrak{P}} \mathrm{inv}_{N_{\mathfrak{P}'}|L_{\mathfrak{P}}}(\mathrm{res}_L c)_{\mathfrak{P}} = \sum_{\mathfrak{P}} \mathrm{inv}_{N_{\mathfrak{P}'}|L_{\mathfrak{P}}}(\mathrm{res}_{L_{\mathfrak{P}}} c_{\mathfrak{p}})$$
$$= \sum_{\mathfrak{P}}[L_{\mathfrak{P}} : K_{\mathfrak{p}}] \cdot \mathrm{inv}_{N_{\mathfrak{P}'}|K_{\mathfrak{p}}} c_{\mathfrak{p}} = \sum_{\mathfrak{p}} \sum_{\mathfrak{P}|\mathfrak{p}}[L_{\mathfrak{P}} : K_{\mathfrak{p}}] \cdot \mathrm{inv}_{N_{\mathfrak{P}'}|K_{\mathfrak{p}}} c_{\mathfrak{p}},$$

where $\mathfrak{P}'$ represents any prime of N over $\mathfrak{P}$ and $\mathfrak{p}$ the prime of K lying under $\mathfrak{P}$. If we note that the invariants $\mathrm{inv}_{N_{\mathfrak{P}'}|K_{\mathfrak{p}}} c_{\mathfrak{p}}$ are independent of the choice of the primes $\mathfrak{P}'$ of N lying above $\mathfrak{p}$ (cf. the footnote [14] on p. 138), and that by the fundamental equation of number theory (cf. §1, p. 115) we have

$$\sum_{\mathfrak{P}|\mathfrak{p}}[L_{\mathfrak{P}} : K_{\mathfrak{p}}] = [L : K],$$

then ($\mathfrak{P}'$ a fixed prime of N over $\mathfrak{p}$):

$$\mathrm{inv}_{N|L}(\mathrm{res}_L c) = \sum_{\mathfrak{p}} \left(\sum_{\mathfrak{P}|\mathfrak{p}}[L_{\mathfrak{P}} : K_{\mathfrak{p}}] \right) \cdot \mathrm{inv}_{N_{\mathfrak{P}'}|K_{\mathfrak{p}}} c_{\mathfrak{p}}$$

$$= [L : K] \cdot \sum_{\mathfrak{p}} \mathrm{inv}_{N_{\mathfrak{P}'}|K_{\mathfrak{p}}} c_{\mathfrak{p}}$$

$$= [L : K] \cdot \mathrm{inv}_{N|K} c.$$

Finally, for $c \in H^2(G_{N|L}, I_N)$ it follows from the formulas in (3.3) that

$$\mathrm{inv}_{N|K}(\mathrm{cor}_K c) = \sum_{\mathfrak{p}} \mathrm{inv}_{N_{\mathfrak{P}'}|K_{\mathfrak{p}}}(\mathrm{cor}_K c)_{\mathfrak{p}}$$

$$= \sum_{\mathfrak{p}} \sum_{\mathfrak{P}|\mathfrak{p}} \mathrm{inv}_{N_{\mathfrak{P}'}|K_{\mathfrak{p}}}(\mathrm{cor}_{K_{\mathfrak{p}}} c_{\mathfrak{P}})$$

$$= \sum_{\mathfrak{p}} \sum_{\mathfrak{P}|\mathfrak{p}} \mathrm{inv}_{N_{\mathfrak{P}'}|L_{\mathfrak{P}}} c_{\mathfrak{P}} = \mathrm{inv}_{N|L}(c).$$

Since $H^1(G_{L|K}, I_L) = 1$, it follows that the extensions $L|K$ satisfy with respect to the idèle group I_L and the idèle homomorphism $\mathrm{inv}_{L|K}$ the conditions for a class formation, except for that the homomorphism $\mathrm{inv}_{L|K} :$ $H^2(G_{L|K}, I_L) \to \frac{1}{[L:K]}\mathbb{Z}/\mathbb{Z}$ is not an isomorphism. To makes this an isomorphism, we have to pass from the idèle group I_L to the idèle class group C_L. Before explaining this in detail, we consider abelian extensions and introduce an invariant homomorphism $\mathrm{inv}_{L|K}$, as well as the following symbol.

(5.3) Definition. Let $L|K$ be an abelian extension. If $\mathfrak{a} \in I_K$ with local components $\mathfrak{a}_{\mathfrak{p}} \in K_{\mathfrak{p}}^{\times}$, then we set

$$(\mathfrak{a}, L|K) = \prod_{\mathfrak{p}} (\mathfrak{a}_{\mathfrak{p}}, L_{\mathfrak{P}}|K_{\mathfrak{p}}) \in G_{L|K} .$$

In this definition we have used the following facts: For each prime $\mathfrak{p}$, the symbol $(\mathfrak{a}_\mathfrak{p}, L_\mathfrak{P}|K_\mathfrak{p})$ defines an element of the local abelian Galois group $G_{L_\mathfrak{P}|K_\mathfrak{p}}$, which we always consider as a subgroup of $G_{L|K}$. Hence

$$(\mathfrak{a}_\mathfrak{p}, L_\mathfrak{P}|K_\mathfrak{p}) \in G_{L_\mathfrak{P}|K_\mathfrak{p}} \subseteq G_{L|K}.$$

Since $\mathfrak{a}_\mathfrak{p}$ is a unit for almost all primes $\mathfrak{p}$ and since $L_\mathfrak{P}|K_\mathfrak{p}$ is unramified for almost all $\mathfrak{p}$, we have $(\mathfrak{a}_\mathfrak{p}, L_\mathfrak{P}|K_\mathfrak{p}) = 1$ for almost all $\mathfrak{p}$ by II, (4.4). Thus the product $\prod_\mathfrak{p}(\mathfrak{a}_\mathfrak{p}, L_\mathfrak{P}|K_\mathfrak{p}) \in G_{L|K}$ is well defined, and it is also independent of the order of the factors, since $G_{L|K}$ is abelian. The symbol $(\ , L|K)$ and the invariant mapping $\mathrm{inv}_{L|K}$ are related as follows.

(5.4) Lemma[15]. *Let* $L|K$ *be an abelian extension,* $\mathfrak{a} \in I_K$ *and* $(\mathfrak{a}) = \mathfrak{a} \cdot N_{L|K} I_L \in H^0(G_{L|K}, I_L)$. *If* $\chi \in \chi(G_{L|K}) = H^1(G_{L|K}, \mathbb{Q}/\mathbb{Z})$, *then*

$$\chi(\mathfrak{a}, L|K) = \mathrm{inv}_{L|K}((\mathfrak{a}) \cup \delta\chi) \in \tfrac{1}{[L:K]}\mathbb{Z}/\mathbb{Z}.$$

This is a consequence of the analogous formula which relates the local norm residue symbol $(\ , L_\mathfrak{P}|K_\mathfrak{p})$ with the local invariant map $\mathrm{inv}_{L_\mathfrak{P}|K_\mathfrak{p}}$ (cf. II, (5.11)). If we denote by $\chi_\mathfrak{p}$ the restriction of χ to $G_{L_\mathfrak{P}|K_\mathfrak{p}}$, and by $(\mathfrak{a}_\mathfrak{p}) = \mathfrak{a}_\mathfrak{p} \cdot N_{L_\mathfrak{P}|K_\mathfrak{p}} L_\mathfrak{P}^\times$, then

$$\chi(\mathfrak{a}, L|K) = \sum_\mathfrak{p} \chi_\mathfrak{p}(\mathfrak{a}_\mathfrak{p}, L_\mathfrak{P}|K_\mathfrak{p}) = \sum_\mathfrak{p} \mathrm{inv}_{L_\mathfrak{P}|K_\mathfrak{p}}((\mathfrak{a}_\mathfrak{p}) \cup \delta\chi_\mathfrak{p}).$$

The remark after (3.2) shows that the classes

$$((\mathfrak{a}_\mathfrak{p}) \cup \delta\chi_\mathfrak{p}) \in H^2(G_{L_\mathfrak{P}|K_\mathfrak{p}}, L_\mathfrak{P}^\times)$$

are the local components of $(\mathfrak{a}) \cup \delta\chi \in H^2(G_{L|K}, I_L)$; we only need to note that $\mathfrak{a}_\mathfrak{p} \cdot \delta\chi_\mathfrak{p}(\sigma, \tau)$ (resp. $\mathfrak{a} \cdot \delta\chi(\sigma, \tau)$) is a 2-cocycle of the class $((\mathfrak{a}_\mathfrak{p}) \cup \delta\chi_\mathfrak{p})$ (resp. $((\mathfrak{a})\cup\delta\chi)$) (cf. II, (5.11)). Thus $\chi(\mathfrak{a}, L|K) = \mathrm{inv}_{L|K}((\mathfrak{a})\cup\delta\chi)$ as claimed.

When changing from the idèle invariants to the idèle class invariants, the following theorem is of central importance. From the exact cohomology sequence associated with the exact sequence

$$1 \longrightarrow L^\times \longrightarrow I_L \longrightarrow C_L \longrightarrow 1$$

we see, using $H^1(G_{L|K}, C_L) = 1$, that the induced homomorphism

$$H^2(G_{L|K}, L^\times) \longrightarrow H^2(G_{L|K}, I_L)$$

is injective.

We use this injection to think of $H^2(G_{L|K}, L^\times)$ as a subgroup of $H^2(G_{L|K}, I_L)$, i.e., we view the elements of $H^2(G_{L|K}, L^\times)$ as the idèle cohomology classes that are represented by cocycles with values in the principal idèle group $L^\times$.

(5.5) Theorem. *If* $c \in H^2(G_{L|K}, L^\times)$, *then* $\mathrm{inv}_{L|K} c = 0$.

---

[15] Cf. also II, (1.10).

We will see that apart from purely technical considerations the proof of this theorem is based on the following two facts: the explicit description of the local norm residue symbol and the product formula for algebraic numbers.

Proof. We start with the simple observation that it suffices to consider the case when $K = \mathbb{Q}$ and L is a cyclic cyclotomic extension of $\mathbb{Q}$. In fact, if $c \in H^2(G_{L|K}, L^\times)$, and N is a normal extension of $\mathbb{Q}$ containing L, then

$$c \in H^2(G_{L|K}, L^\times) \subseteq H^2(G_{N|K}, N^\times) \subseteq H^2(G_{N|K}, I_N),$$

$\mathrm{cor}_\mathbb{Q} c \in H^2(G_{N|\mathbb{Q}}, N^\times)$, and $\mathrm{inv}_{L|K} c = \mathrm{inv}_{N|K} c = \mathrm{inv}_{N|\mathbb{Q}}(\mathrm{cor}_\mathbb{Q} c)$ by (5.2). Hence to show $\mathrm{inv}_{L|K} c = 0$ it suffices to consider the case $K = \mathbb{Q}$. Since by (3.6) there exists a cyclic cyclotomic extension $L_0|\mathbb{Q}$ with $c \in H^2(G_{L_0|\mathbb{Q}}, L_0^\times)$, we can even assume that $L|\mathbb{Q}$ itself is a cyclic cyclotomic extension.

Let χ be a generator of the cyclic character group $\chi(G_{L|\mathbb{Q}}) = H^1(G_{L|\mathbb{Q}}, \mathbb{Q}/\mathbb{Z})$. Then $\delta\chi$ is a generator of $H^2(G_{L|\mathbb{Q}}, \mathbb{Z})$, and Tate's Theorem I, (7.3) implies

$$\delta\chi \cup \ : H^0(G_{L|\mathbb{Q}}, L^\times) \longrightarrow H^2(G_{L|\mathbb{Q}}, L^\times) \quad \text{16)}$$

is bijective. Thus each element $c \in H^2(G_{L|\mathbb{Q}}, L^\times)$ has the form $c = (a) \cup \delta\chi$ with $(a) = a \cdot N_{L|\mathbb{Q}} L^\times \in H^0(G_{L|\mathbb{Q}}, L^\times)$, $a \in \mathbb{Q}^\times$. From (5.4) we obtain

$$\mathrm{inv}_{L|\mathbb{Q}} c = \mathrm{inv}_{L|\mathbb{Q}}((a) \cup \delta\chi) = \chi(a, L|\mathbb{Q}),$$

and we need to show that $(a, L|\mathbb{Q}) = \prod_p (a, L_\mathfrak{P}|\mathbb{Q}_p) = 1$. Now L is a cyclotomic extension, i.e., $L \subseteq \mathbb{Q}(\zeta)$ for some root of unity ζ. The automorphism $(a, L|\mathbb{Q})$ is precisely the restriction of $(a, \mathbb{Q}(\zeta)|\mathbb{Q})$ to L; this follows easily from the behavior of the local norm residue symbol $(a, \mathbb{Q}_p(\zeta)|\mathbb{Q}_p)$ when passing to the extension $L_\mathfrak{P}|\mathbb{Q}_p$ (cf. II, (5.10a)). It therefore suffices to show that $(a, \mathbb{Q}(\zeta)|\mathbb{Q}) = 1$ for $a \in \mathbb{Q}^\times$. Now $\mathbb{Q}(\zeta)$ is generated by roots of unity of prime power order and it suffices to show the vanishing of $(a, \mathbb{Q}(\zeta)|\mathbb{Q})$ for these generators, hence we may assume that ζ is a primitive l^n-th root of unity (l a prime number). With this reduction we come to the actual core of the proof.

Let ζ be a primitive l^n-th root of unity; if $l = 2$, we assume $n \geq 2$. If p ranges over the prime numbers and the infinite prime over $p = p_\infty$ of $\mathbb{Q}$, then the $\mathbb{Q}_p(\zeta)|\mathbb{Q}_p$ are the local extensions associated with $\mathbb{Q}(\zeta)|\mathbb{Q}$. The extension $\mathbb{Q}_p(\zeta)|\mathbb{Q}_p$ is unramified for $p \neq l$ and totally ramified for $p = l$ (cf. [21], §5, 3.); if $p = p_\infty$, then $\mathbb{Q}_p(\zeta)|\mathbb{Q}_p$ means the extension $\mathbb{C}|\mathbb{R}$. We have to show:

For each $a \in \mathbb{Q}^\times$, $(a, \mathbb{Q}(\zeta)|\mathbb{Q}) = \prod_p (a, \mathbb{Q}_p(\zeta)|\mathbb{Q}_p) = 1$.

Here it obviously suffices to assume that a is integral. We consider the effect of the local norm residue symbol $(a, \mathbb{Q}_p(\zeta)|\mathbb{Q}_p)$ on the l^n-th roots of unity ζ.

1. For $p \neq l$, $p \neq p_\infty$, we have by II, (4.8)

$$(a, \mathbb{Q}_p(\zeta)|\mathbb{Q}_p)\zeta = \varphi^{v_p(a)}\zeta,$$

---

16) This can also be shown in an elementary way.

where v_p is the valuation on $\mathbb{Q}_p$ and φ is the Frobenius automorphism on $\mathbb{Q}_p(\zeta)|\mathbb{Q}_p$. Since the residue field of $\mathbb{Q}_p$ has p elements, clearly $\varphi\zeta = \zeta^p$, thus

$$(a, \mathbb{Q}_p(\zeta)|\mathbb{Q}_p)\zeta = \zeta^{p^{v_p(a)}}.$$

2. For $p = l$, we obtain from II, (7.16), writing $a = u \cdot p^m = u \cdot p^{v_p(a)}$, u a unit:

$$(a, \mathbb{Q}_p(\zeta)|\mathbb{Q}_p)\zeta = \zeta^r,$$

where r is a natural number which is determined mod p^n by the congruence $r \equiv u^{-1} \equiv a^{-1} \cdot p^{v_p(a)} \bmod p^n$.

3. For $p = p_\infty$ the automorphism $(a, \mathbb{C}|\mathbb{R})$ is either the identity or complex conjugation, depending on whether $a > 0$ or $a < 0$ (cf. II, §5, p. 95). Thus

$$(a, \mathbb{Q}_p(\zeta)|\mathbb{Q}_p)\zeta = \zeta^{\mathrm{sgn}\, a}.$$

Combining all of the above, we obtain

$$(a, \mathbb{Q}(\zeta)|\mathbb{Q})\zeta = \prod_p (a, \mathbb{Q}_p(\zeta)|\mathbb{Q}_p)\zeta = \zeta^{\mathrm{sgn}\, a \cdot \prod_{p \neq l} p^{v_p(a)} \cdot r}.$$

But by the product formula we have for the power appearing on the right side

$$\mathrm{sgn}\, a \cdot \prod_{p \neq l} p^{v_p(a)} \cdot r \equiv \mathrm{sgn}\, a \cdot \prod_{p \neq l} p^{v_p(a)} l^{v_\ell(a)} a^{-1} = \frac{1}{\prod_p |a|_p} = 1 \bmod l^n,$$

therefore $(a, \mathbb{Q}(\zeta)|\mathbb{Q})\zeta = \zeta$, i.e., we have in fact $(a, \mathbb{Q}(\zeta)|\mathbb{Q}) = 1$.

Theorem (5.5) shows that the group $H^2(G_{L|K}, L^\times)$ lies in the kernel of the homomorphism $\mathrm{inv}_{L|K} : H^2(G_{L|K}, I_L) \to \frac{1}{[L:K]}\mathbb{Z}/\mathbb{Z}$. We have to ask further whether or not it is precisely the kernel, and in addition whether or not $\mathrm{inv}_{L|K}$ is a surjective homomorphism. For the cyclic case we have:

(5.6) Proposition. *If $L|K$ is a cyclic extension, then the sequence*

$$1 \longrightarrow H^2(G_{L|K}, L^\times) \longrightarrow H^2(G_{L|K}, I_L) \xrightarrow{\mathrm{inv}_{L|K}} \frac{1}{[L:K]}\mathbb{Z}/\mathbb{Z} \longrightarrow 0$$

is exact.

Proof. a) To show $\mathrm{inv}_{L|K}$ is surjective, we assume first $[L : K]$ is a prime power p^r. Because $\frac{1}{[L:K]} + \mathbb{Z}$ generates $\frac{1}{[L:K]}\mathbb{Z}/\mathbb{Z}$, it suffices to find an element $c \in H^2(G_{L|K}, I_L)$ with $\mathrm{inv}_{L|K} c = \frac{1}{[L:K]} + \mathbb{Z}$. We use the decomposition

$$H^2(G_{L|K}, I_L) \cong \bigoplus_{\mathfrak{p}} H^2(G_{L_{\mathfrak{P}}|K_{\mathfrak{p}}}, L_{\mathfrak{P}}^\times)$$

and determine c by its local components $c_{\mathfrak{p}} \in H^2(G_{L_{\mathfrak{P}}|K_{\mathfrak{p}}}, L_{\mathfrak{P}}^\times)$. Since $L|K$ is cyclic of prime power degree, it follows from (4.3) that K contains a prime $\mathfrak{p}_0$ which is inert in L. Since $\mathfrak{p}_0$ is inert, we have $[L_{\mathfrak{P}_0} : K_{\mathfrak{p}_0}] = [L : K]$, $(\mathfrak{P}_0|\mathfrak{p}_0)$, and local class field theory yields an element $c_{\mathfrak{p}_0} \in H^2(G_{L_{\mathfrak{P}_0}|K_{\mathfrak{p}_0}}, L_{\mathfrak{P}_0}^\times)$ with

$\mathrm{inv}_{L_{\mathfrak{P}_0}|K_{\mathfrak{p}_0}} c_{\mathfrak{p}_0} = \frac{1}{[L_{\mathfrak{P}_0}:K_{\mathfrak{p}_0}]} + \mathbb{Z} = \frac{1}{[L:K]} + \mathbb{Z}$. Now if c is the element in $H^2(G_{L|K}, I_L)$ that is determined by the local components

$$\ldots, 1, 1, 1, c_{\mathfrak{p}_0}, 1, 1, 1, \ldots,$$

then

$$\mathrm{inv}_{L|K} c = \sum_{\mathfrak{p}} \mathrm{inv}_{L_{\mathfrak{P}}|K_{\mathfrak{p}}} c_{\mathfrak{p}} = \mathrm{inv}_{L_{\mathfrak{P}_0}|K_{\mathfrak{p}_0}} c_{\mathfrak{p}_0} = \frac{1}{[L:K]} + \mathbb{Z}.$$

That $\mathrm{inv}_{L|K}$ is also surjective in the general case $[L : K] = n = p_1^{r_1} \cdots p_s^{r_s}$ follows easily from this. For every $i = 1, \ldots, s$ there obviously exists a cyclic intermediate field L_i of degree $[L_i : K] = p_i^{r_i}$. Consider the decomposition

$$\frac{1}{n} = \frac{n_1}{p_1^{r_1}} + \cdots + \frac{n_s}{p_s^{r_s}}$$

into partial fraction. By the previous case there is a $c_i \in H^2(G_{L_i|K}, I_{L_i})$ with

$$\mathrm{inv}_{L_i|K} c_i = \mathrm{inv}_{L|K} c_i = \frac{n_i}{p_i^{r_i}} + \mathbb{Z}.$$

Thus if we set

$$c = c_1 \cdots c_s \in H^2(G_{L|K}, I_L),$$

then

$$\mathrm{inv}_{L|K} c = \sum_{i=1}^{s} \mathrm{inv}_{L|K} c_i = \sum_{i=1}^{s} \frac{n_i}{p_i^{r_i}} + \mathbb{Z} = \frac{1}{n} + \mathbb{Z},$$

which shows that $\mathrm{inv}_{L|K}$ is surjective for any cyclic extension.

b) By (5.5) $H^2(G_{L|K}, L^\times)$ is contained in the kernel of $\mathrm{inv}_{L|K}$. To show that the group $H^2(G_{L|K}, L^\times)$ in fact equals the kernel of $\mathrm{inv}_{L|K}$ we use a simple argument involving the orders of these groups. Since the map $\mathrm{inv}_{L|K}$ is surjective, we only need to show that the order of the factor group

$$H^2(G_{L|K}, I_L)/H^2(G_{L|K}, L^\times)$$

is at most the order of $\frac{1}{[L:K]}\mathbb{Z}/\mathbb{Z}$, i.e., the degree of $[L : K]$. Using the sequence

$$1 \longrightarrow L^\times \longrightarrow I_L \longrightarrow C_L \longrightarrow 1$$

we obtain, using that $H^1(G_{L|K}, C_L) = 1$, the exact cohomology sequence

$$1 \longrightarrow H^2(G_{L|K}, L^\times) \longrightarrow H^2(G_{L|K}, I_L) \longrightarrow H^2(G_{L|K}, C_L).$$

Therefore the order of $H^2(G_{L|K}, I_L)/H^2(G_{L|K}, L^\times)$ divides the order of $H^2(G_{L|K}, C_L)$. By (4.9) $H^2(G_{L|K}, C_L)$ divides $[L : K]$, and we are done.

For the following it would be very convenient if we could show that $\mathrm{inv}_{L|K}$ is a surjective homomorphism in general. Unfortunately this is false. In order for every element of $\frac{1}{[L:K]}\mathbb{Z}/\mathbb{Z}$ to be in the image of the invariant map, it is necessary to enlarge the field L by forming the compositum with a cyclic extension. For technical reasons it is best to let L range over all normal extensions of K and to consider the union

$$H^2(G_{\Omega|K}, I_\Omega) = \bigcup_L H^2(G_{L|K}, I_L)$$

(cf. remark after (3.5)). If $N \supseteq L \supseteq K$ are two normal extensions of K, then

$$H^2(G_{L|K}, I_L) \subseteq H^2(G_{N|K}, I_N),$$

and since by (5.2) the invariant map can be extended from $H^2(G_{L|K}, I_L)$ to $H^2(G_{N|K}, I_N)$, we obtain a homomorphism

$$\mathrm{inv}_K : H^2(G_{\Omega|K}, I_\Omega) \longrightarrow \mathbb{Q}/\mathbb{Z},$$

whose restriction to $H^2(G_{L|K}, I_L) \subseteq H^2(G_{\Omega|K}, I_\Omega)$ coincides with the initial invariant map $\mathrm{inv}_{L|K}$. If we take into account that for each positive integer m there is a cyclic extension $L|K$ with $m \mid [L:K]$ (for example, by (3.7)), we see that $\mathbb{Q}/\mathbb{Z}$ is already covered by the groups $\frac{1}{[L:K]}\mathbb{Z}/\mathbb{Z}$ coming from cyclic extension $L|K$. Now the map $\mathrm{inv}_{L|K}$ is surjective in the cyclic case, thus we obtain for the invariant map inv_K defined above the following

(5.7) Theorem. *The homomorphism*

$$\mathrm{inv}_K : H^2(G_{\Omega|K}, I_\Omega) \longrightarrow \mathbb{Q}/\mathbb{Z}$$

is surjective.

The results obtained so far can be summarized to give a theorem about the Brauer group of an algebraic number field K and its completion $K_\mathfrak{p}$. Recall that we defined the Brauer group $Br(K)$ of a field K in II, §2 as the union (more precisely the direct limit)

$$Br(K) = \bigcup_L H^2(G_{L|K}, L^\times),$$

where L runs through all finite Galois extensions of K. If K is an algebraic number field, then we choose over each prime $\mathfrak{p}$ of K a fixed valuation of the algebraic closure Ω; each such valuation in turn determines in every finite extension $L|K$ a prime $\mathfrak{P}$ above $\mathfrak{p}$. Then we have

$$Br(K_\mathfrak{p}) = \bigcup_L H^2(G_{L_\mathfrak{P}|K_\mathfrak{p}}, L_\mathfrak{P}^\times).$$

From the homomorphisms

$$H^2(G_{L|K}, L^\times) \longrightarrow H^2(G_{L|K}, I_L) \cong \bigoplus_\mathfrak{p} H^2(G_{L_\mathfrak{P}|K_\mathfrak{p}}, L_\mathfrak{P}^\times) \xrightarrow{\mathrm{inv}_{L|K}} \frac{1}{[L:K]}\mathbb{Z}/\mathbb{Z}$$

we obtain by passing to the direct limit (i.e., in this case, the union) the canonical homomorphism

$$Br(K) \longrightarrow H^2(G_{\Omega|K}, I_\Omega) \cong \bigoplus_\mathfrak{p} Br(K_\mathfrak{p}) \xrightarrow{\mathrm{inv}_K} \mathbb{Q}/\mathbb{Z},$$

where inv_K is the sum of the local invariant maps $\mathrm{inv}_{K_\mathfrak{p}} : Br(K_\mathfrak{p}) \longrightarrow \mathbb{Q}/\mathbb{Z}$ (cf. II, §1, p. 68 and II, (5.4)).

We now have **Hasse's Main Theorem on the Theory of Algebras**:

(5.8) Theorem. *For every finite algebraic number field K one has the canonical exact sequence*

$$1 \longrightarrow Br(K) \longrightarrow \bigoplus_{\mathfrak{p}} Br(K_{\mathfrak{p}}) \xrightarrow{\mathrm{inv}_K} \mathbb{Q}/\mathbb{Z} \longrightarrow 0 .$$

Proof. The groups $Br(K)$, $\bigoplus_{\mathfrak{p}} Br(K_{\mathfrak{p}})$ ($\cong H^2(G_{\Omega|K}, I_\Omega)$), and $\mathbb{Q}/\mathbb{Z}$ are, by (3.6) and the remark after (5.7), the union of the groups $H^2(G_{L|K}, L^\times)$, $\bigoplus_{\mathfrak{p}} H^2(G_{L_{\mathfrak{P}}|K_{\mathfrak{p}}}, L_{\mathfrak{P}}^\times)$ ($\cong H^2(G_{L|K}, I_L)$), and $\frac{1}{[L:K]}\mathbb{Z}/\mathbb{Z}$, respectively, provided $L|K$ runs through all the cyclic extensions. But for such a cyclic extension $L|K$ we have by (5.6) the exact sequence

$$1 \longrightarrow H^2(G_{L|K}, L^\times) \longrightarrow H^2(G_{L|K}, I_L) \xrightarrow{\mathrm{inv}_{L|K}} \frac{1}{[L:K]}\mathbb{Z}/\mathbb{Z} \longrightarrow 0,$$

from which the theorem follows immediately.

§ 6. The Reciprocity Law

Having studied the idèle invariants in the previous section, we now want to construct invariants for the elements of the groups $H^2(G_{L|K}, C_L)$. We start with the following observations:

If $L|K$ is a normal extension, then we obtain from the exact sequence

$$1 \longrightarrow L^\times \longrightarrow I_L \longrightarrow C_L \longrightarrow 1,$$

using $H^1(G_{L|K}, C_L) = 1 = H^3(G_{L|K}, I_L) = 1$, the exact cohomology sequence

$$1 \longrightarrow H^2(G_{L|K}, L^\times) \longrightarrow H^2(G_{L|K}, I_L) \xrightarrow{j} H^2(G_{L|K}, C_L)$$

$$\xrightarrow{\delta} H^3(G_{L|K}, L^\times) \longrightarrow 1.$$

If $\overline{c} \in H^2(G_{L|K}, C_L)$ and $c \in H^2(G_{L|K}, I_L)$ is such that $\overline{c} = jc$, then we set

$$\mathrm{inv}_{L|K}\overline{c} = \mathrm{inv}_{L|K}c \in \frac{1}{[L:K]}\mathbb{Z}/\mathbb{Z}.$$

This definition is independent of the choice of the preimage $c \in H^2(G_{L|K}, I_L)$, because two such preimages differ only by an element in $H^2(G_{L|K}, L^\times)$, which by (5.5) has invariant 0. Of course, this only works if the element $\overline{c} \in H^2(G_{L|K}, C_L)$ lies in the image of the homomorphism j. But the map j is in general not surjective; in fact, j being surjective would be equivalent to the group $H^3(G_{L|K}, L^\times)$ being trivial, however, this group is not equal to 1 in general (cf. [2], Ch. 7, Th. 12). Nevertheless, we can show that at least in the cyclic case the map j is surjective.

(6.1) Proposition. *If $L|K$ is a cyclic extension, then the homomorphism*

$$H^2(G_{L|K}, I_L) \xrightarrow{\ j\ } H^2(G_{L|K}, C_L)$$

is surjective.

Proof. In $L|K$ is cyclic, then $H^3(G_{L|K}, L^\times) \cong H^1(G_{L|K}, L^\times) = 1$ (cf. II, (2.2)).

In order to define an invariant map for arbitrary normal extensions $L|K$, we proceed in a similar way as for the groups $H^2(G_{L|K}, I_L)$ with respect to the invariant map for idèles at the end of the previous section.

First we note that the homomorphism

$$H^2(G_{L|K}, I_L) \xrightarrow{\ j\ } H^2(G_{L|K}, C_L)$$

commutes with the maps inf and res; i.e., if $N \supseteq L \supseteq K$ are two normal extensions of K, then we have

$$j \circ \mathrm{inf}_N = \mathrm{inf}_N \circ j \quad \text{and} \quad j \circ \mathrm{res}_L = \mathrm{res}_L \circ j \,,$$

where in the last formula we only need to assume that $N|K$ is normal. For simplicity we set

(6.2) Definition. $H^q(L|K) = H^q(G_{L|K}, C_L)$.

Because $H^1(L|K) = 1$ (cf. (4.7)), the extensions $L|K$ form a field formation in the sense of II, §1 with respect to the idèle group C_L as the module. To simplify things we will, as for the idèle cohomology groups in §5 and generally for every field formation, interpret the injective inflation maps

$$H^2(L|K) \xrightarrow{\ \mathrm{inf}\ } H^2(N|K) \qquad (N \supseteq L \supseteq K)$$

as inclusions. More precisely, this means that we form the direct limit

$$H^2(\Omega|K) = \varinjlim_L H^2(L|K) \quad {}^{17)}$$

where L ranges over all finite normal extensions of K. We view the groups $H^2(L|K)$ as being embedded in $H^2(\Omega|K)$ via the inflation maps. Thinking of the $H^2(L|K)$ as subgroups of $H^2(\Omega|K)$, we have

$$H^2(\Omega|K) = \bigcup_L H^2(L|K).$$

Hence if $N \supseteq L \supseteq K$ are two normal extensions, then we have inclusions

$$H^2(L|K) \subseteq H^2(N|K) \subseteq H^2(\Omega|K),$$

---

[17] Here Ω denotes again the field of all algebraic numbers, and the same convention as in footnote [9] for the group $H^2(G_{\Omega|K}, I_\Omega)$ applies to the group $H^2(\Omega|K)$.

where, according to this interpretation, an element $\bar{c} \in H^2(N|K)$ comes from an element of $H^2(L|K)$ if and only if it is the inflation of an element of $H^2(L|K)$.

Now the crucial input is the following theorem which plays a role similar to one of Theorem (II, 5.2) in the local theory.

(6.3) Theorem. *If $L|K$ is a normal extension and $L'|K$ a cyclic extension of equal degree $[L' : K] = [L : K]$, then*

$$H^2(L'|K) = H^2(L|K) \subseteq H^2(\Omega|K).$$

Since for every positive integer m there is a cyclic extension $L|K$ of degree m (see (3.7), for instance), this theorem has the following

(6.4) Corollary. $H^2(\Omega|K) = \bigcup_{L|K \text{ cyclic}} H^2(L|K)$.

Proof of (6.3). We first show that $H^2(L'|K) \subseteq H^2(L|K)$. If $N = L \cdot L'$ is the compositum of L and L', then a simple group theoretic argument shows that if the extension $L'|K$ is cyclic, then the extension $N|L$ is also cyclic. Now let $\bar{c} \in H^2(L'|K) \subseteq H^2(N|K)$. Because of the exact sequence

$$1 \longrightarrow H^2(L|K) \longrightarrow H^2(N|K) \xrightarrow{\text{res}_L} H^2(N|L),$$

an element $\bar{c} \in H^2(N|K)$ is an element of $H^2(L|K)$ if and only if $\text{res}_L \bar{c} = 1$. To show this, we use the idèle invariants. By (6.1) the homomorphism

$$H^2(G_{L'|K}, I_{L'}) \xrightarrow{j} H^2(L'|K)$$

is surjective, so that $\bar{c} = jc$, $c \in H^2(G_{L'|K}, I_{L'}) \subseteq H^2(G_{N|K}, I_N)$. From the remarks made above, we know that the map j commutes with inflation (interpreted here as inclusion) and with restriction, hence we have the formulas

$$\text{res}_L \bar{c} = \text{res}_L(jc) = j\text{res}_L c.$$

Thus $\text{res}_L \bar{c} = 1$ if and only if $\text{res}_L c$ lies in the kernel of j, and therefore in $H^2(G_{N|L}, N^\times)$. Since $N|L$ is cyclic, this holds by (5.6) if and only if $\text{inv}_{N|L}(\text{res}_L c) = 0$, and this last statement now follows from

$$\text{inv}_{N|L}(\text{res}_L c) = [L : K] \cdot \text{inv}_{N|K} c = [L' : K] \cdot \text{inv}_{L'|K} c = 0.$$

Therefore $H^2(L'|K) \subseteq H^2(L|K)$.

To show that the above inequality is in fact an equality we consider orders. Because $H^1(L'|K) = 1$ and $H^3(G_{L'|K}, L'^\times) \cong H^1(G_{L'|K}, L'^\times) = 1$, we obtain the exact cohomology sequence

$$1 \longrightarrow H^2(G_{L'|K}, L'^\times) \longrightarrow H^2(G_{L'|K}, I_{L'}) \longrightarrow H^2(L'|K) \longrightarrow 1,$$

where $|H^2(L'|K)| = [L' : K] = [L : K]$ by (5.6). On the other hand, $|H^2(L|K)|$ divides the degree $[L : K]$ by (4.9), hence $H^2(L'|K) = H^2(L|K)$.

Let $N \supseteq L \supseteq K$ be two normal extensions. Because the map

$$H^2(G_{L|K}, I_L) \xrightarrow{j} H^2(L|K)$$

is compatible with inflation, it can be extended to a canonical homomorphism

$$H^2(G_{N|K}, I_N) \xrightarrow{j} H^2(N|K).$$

Thus we obtain a homomorphism

$$H^2(G_{\Omega|K}, I_\Omega) \xrightarrow{j} H^2(\Omega|K)$$

whose restriction to the groups $H^2(G_{L|K}, I_L)$ are the initial homomorphisms $H^2(G_{L|K}, I_L) \to H^2(L|K)$. If these are not surjective, then we still have the

(6.5) Theorem. *The homomorphism*

$$H^2(G_{\Omega|K}, I_\Omega) \xrightarrow{j} H^2(\Omega|K)$$

is surjective.

Proof. If $\bar{c} \in H^2(\Omega|K)$, then it follows from (6.4) that there is a cyclic extension $L|K$ such that $\bar{c} \in H^2(L|K)$. Since for a cyclic extension the map

$$H^2(G_{L|K}, I_L) \xrightarrow{j} H^2(L|K)$$

is surjective by (6.1), $\bar{c} = jc$ for some $c \in H^2(G_{L|K}, I_L) \subseteq H^2(G_{\Omega|K}, I_\Omega)$.

Given this theorem, it is easy to obtain class invariants for the elements of $H^2(\Omega|K) = \bigcup_L H^2(L|K)$ from the invariant map of the idèle cohomology classes. From the homomorphism

$$\mathrm{inv}_K : H^2(G_{\Omega|K}, I_\Omega) \longrightarrow \mathbb{Q}/\mathbb{Z},$$

which is surjective by (5.7), we in fact come to the following

(6.6) Definition. *If $\bar{c} \in H^2(\Omega|K)$ and $\bar{c} = jc$, $c \in H^2(G_{\Omega|K}, I_\Omega)$, then define*

$$\mathrm{inv}_K \bar{c} = \mathrm{inv}_K c \in \mathbb{Q}/\mathbb{Z}.$$

Of course we have to convince ourselves that this definition is independent of the choice of the preimage $c \in H^2(G_{\Omega|K}, I_\Omega)$ of $\bar{c}$. But if c' is another element in $H^2(G_{\Omega|K}, I_\Omega)$ with $\bar{c} = jc'$, then $c, c' \in H^2(G_{L|K}, I_L) \subseteq H^2(G_{\Omega|K}, I_\Omega)$ for a sufficiently large normal extension $L|K$, where we may assume that this extension is so large that $\bar{c} \in H^2(L|K)$. Because $\bar{c} = jc = jc'$, c and c' differ only by an element in the kernel of the mapping $j : H^2(G_{L|K}, I_L) \longrightarrow H^2(L|K)$, and thus by an element of $H^2(G_{L|K}, L^\times)$, which has invariant 0 by (5.5).

We remark that in the above definition Theorem (5.5) plays an essential role, in fact one may consider this result an important step in the direction of the reciprocity law. From (6.6) we obtain a homomorphism

$$\mathrm{inv}_K : H^2(\Omega|K) \longrightarrow \mathbb{Q}/\mathbb{Z}.$$

The restriction of inv_K to the group $H^2(L|K)$ coming from a finite normal extension $L|K$ yields a homomorphism

$$\mathrm{inv}_{L|K} : H^2(L|K) \longrightarrow \tfrac{1}{[L:K]}\mathbb{Z}/\mathbb{Z},$$

because the orders of the elements in $H^2(L|K)$ divide the degree $[L:K]$ (cf. I, (3.16)), and consequently are mapped to the only subgroup $\tfrac{1}{[L:K]}\mathbb{Z}/\mathbb{Z}$ of $\mathbb{Q}/\mathbb{Z}$ of order $[L:K]$.

We briefly recall the construction of the map $\mathrm{inv}_{L|K} : H^2(L|K) \to \tfrac{1}{[L:K]}\mathbb{Z}/\mathbb{Z}$:

If $\overline{c} \in H^2(L|K)$, then we obtain the invariant $\mathrm{inv}_{L|K}\overline{c}$ by choosing a cyclic extension $L'|K$ of equal degree $[L':K] = [L:K]$, so that by (6.3) $H^2(L'|K) = H^2(L|K)$ ((6.3)); in particular, $\overline{c} \in H^2(L'|K)$. In this cyclic case we have by (6.1) an idèle cohomology class $c \in H^2(G_{L'|K}, I_{L'})$ with $\overline{c} = jc$, and obtain so

$$\mathrm{inv}_{L|K}\overline{c} = \mathrm{inv}_{L'|K}\overline{c} = \mathrm{inv}_{L'|K}c = \sum_{\mathfrak{p}} \mathrm{inv}_{L'_{\mathfrak{P}}|K_{\mathfrak{p}}} c_{\mathfrak{p}} \in \frac{1}{[L:K]}\mathbb{Z}/\mathbb{Z}.$$

The detour using cyclic extensions, which we have described by introducing the groups $H^2(G_{\Omega|K}, I_{\Omega})$ and $H^2(\Omega|K)$ (and interpreting inflation as inclusion) is necessary, because in general the map

$$H^2(G_{L|K}, I_L) \xrightarrow{\ j\ } H^2(L|K)$$

is not surjective. However, for the elements in the image of j we immediately obtain from Definition (6.6) the simple

(6.7) Proposition. *If* $\overline{c} = jc$, $\overline{c} \in H^2(L|K)$, $c \in H^2(G_{L|K}, I_L)$, *then*

$$\mathrm{inv}_{L|K}\overline{c} = \mathrm{inv}_{L|K}c.$$

(6.8) Theorem. *The invariant maps*

$$\mathrm{inv}_K : H^2(\Omega|K) \longrightarrow \mathbb{Q}/\mathbb{Z}$$

and

$$\mathrm{inv}_{L|K} : H^2(L|K) \longrightarrow \tfrac{1}{[L:K]}\mathbb{Z}/\mathbb{Z}$$

are isomorphisms.

Proof. It suffices to verify that $\mathrm{inv}_{L|K}$ is bijective. Let $L'|K$ be a cyclic extension of degree $[L':K] = [L:K]$, so that $H^2(L'|K) = H^2(L|K)$. If $\alpha \in \tfrac{1}{[L:K]}\mathbb{Z}/\mathbb{Z}$, then by (5.6) there is a $c \in H^2(G_{L'|K}, I_{L'})$ with $\mathrm{inv}_{L'|K}c = \alpha$. Set

$\bar{c} = jc \in H^2(L'|K) = H^2(L|K)$. Then $\mathrm{inv}_{L|K}\bar{c} = \mathrm{inv}_{L'|K}\bar{c} = \mathrm{inv}_{L'|K}c = \alpha$, i.e., $\mathrm{inv}_{L|K}$ is surjective.

That $\mathrm{inv}_{L|K}$ is bijective follows now easily from the fact that the order of $H^2(L|K)$ is a divisor of the degree $[L : K]$ (cf. (4.9)), and therefore a divisor of the order of $\frac{1}{[L:K]}\mathbb{Z}/\mathbb{Z}$.

We now come to the main theorem of class field theory. Let K_0 be a fixed algebraic number field, Ω the field of all algebraic numbers, and $G = G_{\Omega|K_0}$ the Galois group of $\Omega|K_0$. We form the union $C_\Omega = \bigcup_K C_K$, where K runs through all finite extensions of K_0 [18]. Then C_Ω is canonically a G-module: If $\bar{c} \in C_\Omega$, say $\bar{c} \in C_L$ for an appropriate finite normal extension $L|K_0$, we set

$$\sigma\bar{c} = \sigma\big|_L \bar{c} \in C_L \subseteq C_\Omega \quad (\sigma \in G).$$

The pair (G, C_Ω) is obviously a formation in the sense of II, §1, and the fundamental result of all our constructions is the following

(6.9) Theorem. *The formation (G, C_Ω) is a class formation with respect to the invariant map introduced in (6.6).*

For the proof we have to verify the axioms in II, §1, (1.3).

Axiom I: $H^1(L|K) = 1$ for every normal extension $L|K$ of each finite extension field of K_0 (cf. (4.7)).

Axiom II: For every normal extension $L|K$ of each finite extension field of K_0 we have by (6.8) the isomorphism

$$\mathrm{inv}_{L|K} : H^2(L|K) \longrightarrow \tfrac{1}{[L:K]}\mathbb{Z}/\mathbb{Z}.$$

a) If $N \supseteq L \supseteq K$ are two normal extensions and $\bar{c} \in H^2(L|K)$, then $\bar{c} \in H^2(N|K)$, and

$$\mathrm{inv}_{N|K}\bar{c} = \mathrm{inv}_{L|K}\bar{c},$$

since $\mathrm{inv}_{N|K}$, and $\mathrm{inv}_{L|K}$ are defined as the restrictions of inv_K to $H^2(N|K)$ and $H^2(L|K) \subseteq H^2(N|K)$ respectively (cf. p. 150).

b) Let $N \supseteq L \supseteq K$ be two extensions fields of K with $N|K$ normal. If $\bar{c} \in H^2(N|K)$, then $\mathrm{res}_L\bar{c} \in H^2(N|L)$. For the proof of the formula

$$\mathrm{inv}_{N|L}(\mathrm{res}_L\bar{c}) = [L : K] \cdot \mathrm{inv}_{N|K}\bar{c}$$

we use the analogous formula for the idèle invariants (cf. (5.2)). By (6.5) there is a $c \in H^2(G_{\Omega|K}, I_\Omega)$ with $jc = \bar{c}$, where we can assume that there is a normal extension $M|K$ containing N, $M \supseteq N \supseteq L \supseteq K$, such that

---

[18] More precisely, we should write $C_\Omega = \varinjlim C_K$. Nevertheless, we think of C_K as embedded in this direct limit and interpret C_Ω as the union of the C_K.

$c \in H^2(G_{M|K}, I_M)$. From the formulas in (5.2), and using the convention that the inflation maps are to be interpreted as inclusions, we have by (6.7)

$$\text{inv}_{N|L}(\text{res}_L \bar{c}) = \text{inv}_{M|L}(\text{res}_L jc) = \text{inv}_{M|L}(j\text{res}_L c) = \text{inv}_{M|L}(\text{res}_L c)$$
$$= [L:K] \cdot \text{inv}_{M|K} c = [L:K] \cdot \text{inv}_{M|K} jc = [L:K] \cdot \text{inv}_{N|K} \bar{c}.$$

Because of this theorem we can now apply the entire abstract theory of class formations to the case of algebraic number fields. If we again denote by

$$u_{L|K} \in H^2(L|K)$$

the **fundamental class** of the normal extension $L|K$, which is uniquely determined by the formula $\text{inv}_{L|K} u_{L|K} = \frac{1}{[L:K]} + \mathbb{Z}$, then we have the general

(6.10) Theorem. *The homomorphism cup product with the fundamental class*

$$u_{L|K} \cup : H^q(G_{L|K}, \mathbb{Z}) \longrightarrow H^{q+2}(L|K)$$

is bijective.

From this, together with II, (1.8) we immediately obtain the

(6.11) Corollary. $H^3(L|K) = 1$ and $H^4(L|K) \cong \chi(G_{L|K})$.

For the case $q = -2$ Theorem (6.10) yields **Artin's Reciprocity Law**:

(6.12) Theorem. *The map cup product with the fundamental class*

$$G_{L|K}^{\text{ab}} \cong H^{-2}(G_{L|K}, \mathbb{Z}) \xrightarrow{u_{L|K} \cup} H^0(L|K) = C_K/N_{L|K} C_L$$

*yields a canonical isomorphism, i.e., the **reciprocity map** between the abelianization $G_{L|K}^{\text{ab}}$ of the Galois group $G_{L|K}$ and the norm residue group $C_K/N_{L|K} C_L$ of the idèle group C_K*

$$\theta_{L|K} : G_{L|K}^{\text{ab}} \longrightarrow C_K/N_{L|K} C_L.$$

The inverse of the reciprocity map $\theta_{L|K}$ is induced from the homomorphism $(\ ,L|K) : C_K \to G_{L|K}^{\text{ab}}$ with kernel $N_{L|K} C_L$, the **norm residue symbol**.

(6.13) Theorem. *The following sequence is exact*

$$1 \longrightarrow N_{L|K} C_L \longrightarrow C_K \xrightarrow{(\ ,L|K)} G_{L|K}^{\text{ab}} \longrightarrow 1.$$

Because the invariant map is compatible with inflation (inclusion) and restriction, we see that the norm residue symbol behaves with respect to varying field extensions as follows (cf. II, (1.11)):

Let $N \supseteq L \supseteq K$ be two extensions with $N|K$ is normal. Then the following diagrams are commutative:

a)

$$
\begin{array}{ccc}
C_K & \xrightarrow{(\ ,N|K)} & G^{\mathrm{ab}}_{N|K} \\
\mathrm{id} \downarrow & & \downarrow \pi \\
C_K & \xrightarrow{(\ ,L|K)} & G^{\mathrm{ab}}_{L|K}
\end{array}
$$

Hence $(\bar{a}, L|K) = \pi(\bar{a}, N|K) \in G^{\mathrm{ab}}_{L|K}$, for $\bar{a} \in C_K$, if $L|K$ is also normal (in addition to $N|K$). Here π is the canonical projection of $G^{\mathrm{ab}}_{N|K}$ onto $G^{\mathrm{ab}}_{L|K}$.

b)

$$
\begin{array}{ccc}
C_K & \xrightarrow{(\ ,N|K)} & G^{\mathrm{ab}}_{N|K} \\
\mathrm{incl} \downarrow & & \downarrow \mathrm{Ver} \\
C_L & \xrightarrow{(\ ,N|L)} & G^{\mathrm{ab}}_{N|L}
\end{array}
$$

Hence $(\bar{a}, N|L) = \mathrm{Ver}(\bar{a}, N|K) \in G^{\mathrm{ab}}_{N|L}$ for $\bar{a} \in C_K$. Recall that the Verlagerung (transfer) Ver is induced by restriction

$$
G^{\mathrm{ab}}_{N|K} \cong H^{-2}(G_{N|K}, \mathbb{Z}) \xrightarrow{\mathrm{res}} H^{-2}(G_{N|L}, \mathbb{Z}) \cong G^{\mathrm{ab}}_{N|L}.
$$

c)

$$
\begin{array}{ccc}
C_L & \xrightarrow{(\ ,N|L)} & G^{\mathrm{ab}}_{N|L} \\
N_{L|K} \downarrow & & \downarrow \kappa \\
C_K & \xrightarrow{(\ ,N|K)} & G^{\mathrm{ab}}_{N|K}
\end{array}
$$

Hence $(N_{L|K}\bar{a}, N|K) = \kappa(\bar{a}, N|L) \in G^{\mathrm{ab}}_{N|K}$ for $\bar{a} \in C_L$, where κ is the canonical homomorphism from $G^{\mathrm{ab}}_{N|L}$ into $G^{\mathrm{ab}}_{N|K}$.

d)

$$
\begin{array}{ccc}
C_K & \xrightarrow{(\ ,N|K)} & G^{\mathrm{ab}}_{N|K} \\
\sigma \downarrow & & \downarrow \sigma^* \\
C_{\sigma K} & \xrightarrow{(\ ,\sigma N|\sigma K)} & G^{\mathrm{ab}}_{\sigma N|\sigma K}
\end{array}
$$

Hence $(\sigma\bar{a}, \sigma N|\sigma K) = \sigma(\bar{a}, N|K)\sigma^{-1}$ for $\bar{a} \in C_K$, where for $\sigma \in G$ the maps $C_K \xrightarrow{\sigma} C_{\sigma K}$ and $G^{\mathrm{ab}}_{N|K} \xrightarrow{\sigma^*} G^{\mathrm{ab}}_{\sigma N|\sigma K}$ are induced by $\bar{a} \mapsto \sigma\bar{a}$ and $\tau \mapsto \sigma\tau\sigma^{-1}$.

In case $N|K$ is abelian, the homomorphisms between the Galois groups in these diagrams are, with the exception of the Verlagerung, the obvious ones.

If $N|K$ is abelian, i.e., if the Galois groups coincide with their abelianizations, then we have in a) the quotient map, in c) the inclusion, and in d) the isomorphism induced by conjugation.

A subgroup I of the idèle class group C_K of a number field K is called a **norm group** if there is a normal extension $L|K$ with $I = N_{L|K}C_L$. By II, (1.14) we have

(6.14) Theorem. *The map*

$$L \longmapsto I_L = N_{L|K}C_L \subseteq C_K$$

yields an inclusion reversing isomorphism between the abelian extensions $L|K$ and the lattice of norm groups I_L of C_K. Therefore

$$I_{L_1 \cdot L_2} = I_{L_1} \cap I_{L_2} \quad \text{and} \quad I_{L_1 \cap L_2} = I_{L_1} \cdot I_{L_2} .$$

*Every subgroup which contains a norm group is again a norm group. If L and I correspond to each other, then L is called the **class field** associated with I.*

This theorem says that the structure of the abelian extensions of K can already be read off of the idèle class group of the base field K. Of course we have to ask if the norm groups can be characterized only by intrinsic information given by the group C_K and independent of the field extensions, similarly to the norm groups in local class field theory which are determined as the closed subgroups of finite index in the multiplicative group of the base field. Such a characterization is presented in the next section.

The reciprocity map $\theta_{L|K}$ from (6.12) and its inverse, the norm residue symbol $(\ ,L|K)$, are defined canonically; however, their explicit description is still too complicated and abstract. We are therefore interested in finding ways to explicitly compute the norm residue symbol.

This is accomplished by the following beautiful theorem which is essentially due to H. HASSE. It connects in a simple way global and local class field theory.

(6.15) Theorem. *Let $L|K$ be an abelian extension and $\bar{\mathfrak{a}} \in C_K$, $\bar{\mathfrak{a}} = \mathfrak{a} \cdot K^\times$, $\mathfrak{a} \in I_K$. Then we have*

$$(\bar{\mathfrak{a}}, L|K) = \prod_{\mathfrak{p}} (\mathfrak{a}_\mathfrak{p}, L_\mathfrak{P}|K_\mathfrak{p}) \in G_{L|K}.$$

Note here that $(\mathfrak{a}_\mathfrak{p}, L_\mathfrak{P}|K_\mathfrak{p}) \in G_{L_\mathfrak{P}|K_\mathfrak{p}} \subseteq G_{L|K}$, and that the components $\mathfrak{a}_\mathfrak{p}$ of the idèle $\mathfrak{a}$ representing $\bar{\mathfrak{a}}$ are units for almost all $\mathfrak{p}$; in particular, since the extensions $L_\mathfrak{P}|K_\mathfrak{p}$ are almost all unramified, the local norm residue symbols are almost all equal to 1.

Since $(\ ,L|K)$ is the **norm residue symbol** of a class formation, we can use Lemma II, (1.10) to prove the theorem: If we denote by $(\overline{\mathfrak{a}}) = \overline{\mathfrak{a}} \cdot N_{L|K} C_L \in H^0(L|K)$, then for every character $\chi \in \chi(G_{L|K}) = H^1(G_{L|K}, \mathbb{Q}/\mathbb{Z})$ we have

$$\chi(\overline{\mathfrak{a}}, L|K) = \mathrm{inv}_{L|K}((\overline{\mathfrak{a}}) \cup \delta\chi).$$

On the other hand, we have already introduced in (5.3) the notation $(\mathfrak{a}, L|K) = \prod_{\mathfrak{p}}(\mathfrak{a}_{\mathfrak{p}}, L_{\mathfrak{P}}|K_{\mathfrak{p}})$ for the product $\prod_{\mathfrak{p}}(\mathfrak{a}_{\mathfrak{p}}, L_{\mathfrak{P}}|K_{\mathfrak{p}})$, and have shown in (5.4) that

$$\chi(\mathfrak{a}, L|K) = \mathrm{inv}_{L|K}((\mathfrak{a}) \cup \delta\chi),$$

where $(\mathfrak{a}) = \mathfrak{a} \cdot N_{L|K} I_L \in H^0(G_{L|K}, I_L)$. The homomorphism

$$H^q(G_{L|K}, I_L) \xrightarrow{j} H^q(G_{L|K}, C_L)$$

maps $(\mathfrak{a}) \in H^0(G_{L|K}, I_L)$ to $(\overline{\mathfrak{a}}) \in H^0(G_{L|K}, C_L)$, and therefore maps $(\mathfrak{a}) \cup \delta\chi \in H^2(G_{L|K}, I_L)$ to $(\overline{\mathfrak{a}}) \cup \delta\chi \in H^2(G_{L|K}, C_L) = H^2(L|K)$, hence

$$j((\mathfrak{a}) \cup \delta\chi) = (\overline{\mathfrak{a}}) \cup \delta\chi \ .$$

With (6.7) we obtain

$$\chi(\overline{\mathfrak{a}}, L|K) = \mathrm{inv}_{L|K}((\overline{\mathfrak{a}}) \cup \delta\chi) = \mathrm{inv}_{L|K}((\mathfrak{a}) \cup \delta\chi) = \chi(\mathfrak{a}, L|K),$$

and since this holds for all characters $\chi \in \chi(G_{L|K})$, it follows that

$$(\overline{\mathfrak{a}}, L|K) = (\mathfrak{a}, L|K) = \prod_{\mathfrak{p}}(\mathfrak{a}_{\mathfrak{p}}, L_{\mathfrak{P}}|K_{\mathfrak{p}}).$$

Hence the global norm residue symbol $(\ ,L|K)$ is determined by the local norm residue symbols $(\ ,L_{\mathfrak{P}}|K_{\mathfrak{p}})$. We will use this observation later when we analyze the reciprocity law for the prime ideal decomposition in abelian extensions.

We close this section with a remark about the **universal norm residue symbol** $(\ ,K)$ (cf. II, §1, p. 76). For every abelian extension we have the homomorphism

$$C_K \xrightarrow{(\ ,L|K)} G_{L|K}.$$

The projective limit

$$G_K^{\mathrm{ab}} = \varprojlim G_{L|K}$$

of the Galois groups $G_{L|K}$ of all (finite) abelian extensions L of K is the Galois group of the maximal abelian extension field A_K of K. For every $\overline{\mathfrak{a}} \in C_K$ we obtain the element

$$(\overline{\mathfrak{a}}, K) = \varprojlim(\overline{\mathfrak{a}}, L|K) \in G_K^{\mathrm{ab}}$$

as the compatible system formed by the elements $(\overline{\mathfrak{a}}, L|K) \in G_{L|K}^{\mathrm{ab}}$. This yields a homomorphism

$$C_K \xrightarrow{(\ ,K)} G_K^{\mathrm{ab}}$$

whose kernel is the intersection of all norm groups (cf. II, (1.15))

$$D_K = \bigcap_L N_{L|K} C_L,$$

and whose image is a dense subgroup of G_K^{ab}.

Finally, we remark without giving details, that the product formula (6.15) yields an analogous product formula for the universal norm residue symbol

$$(\bar{\mathfrak{a}}, K) = \prod_{\mathfrak{p}} (\mathfrak{a}_{\mathfrak{p}}, K_{\mathfrak{p}}),$$

where $(\ ,K_{\mathfrak{p}})$ denotes the universal norm residue symbol of local class field theory, which can be embedded into the group G_K^{ab} (cf. [2], Ch. 7, Cor. 2).

§ 7. The Existence Theorem

By Theorem (6.14) the abelian extensions of a number field K correspond bijectively to the norm groups of C_K. In this section we will characterize these norm groups, similarly to those in local class field theory, as the subgroups of finite index in C_K that are closed with respect to a canonical topology.

The idèle group I_K of an algebraic number field K is the union of the groups $I_K^S = \prod_{\mathfrak{p} \in S} K_{\mathfrak{p}}^{\times} \times \prod_{\mathfrak{p} \notin S} U_{\mathfrak{p}}$, where S ranges over all finite sets of primes of K. The factors $K_{\mathfrak{p}}^{\times}$ and $U_{\mathfrak{p}}$ are equipped with their respective valuation topologies. These induce the Tychonoff topology on the direct product

$$I_K^S = \prod_{\mathfrak{p} \in S} K_{\mathfrak{p}}^{\times} \times \prod_{\mathfrak{p} \notin S} U_{\mathfrak{p}},$$

so that I_K^S becomes a topological group[19]. If $S \subseteq S'$, then $I_K^S \subseteq I_K^{S'}$, and the Tychonoff topology on $I_K^{S'}$ induces the Tychonoff topology on I_K^S. Thus we have a canonical topology on the idèle group $I_K = \bigcup_S I_K^S$, the so-called **idèle topology**. If we want to define the idèle topology directly, we only need to specify a fundamental system of neighborhoods of the identity of I_K, and it is completely obvious that such a fundamental system of neighborhoods is given by the subsets

$$\prod_{\mathfrak{p} \in S} W_{\mathfrak{p}} \times \prod_{\mathfrak{p} \notin S} U_{\mathfrak{p}} \subseteq I_K,$$

where the $W_{\mathfrak{p}} \subseteq K_{\mathfrak{p}}^{\times}$ range over a fundamental system of neighborhoods of the identity of $K_{\mathfrak{p}}^{\times}$, and S over all finite sets of primes of K.

In the following we always consider I_K equipped with this canonical topology. Roughly speaking, two idèles are close to each other when they are close

---

[19] For the theory of topological groups we refer to [9], Ch. III.

componentwise for many primes. The idèle topology is Hausdorff, since the valuation topologies on $K_{\mathfrak{p}}^{\times}$, and therefore the Tychonoff topology on I_K^S are Hausdorff.

(7.1) Proposition. *The idèle group I_K is locally compact.*

Proof. If $W_{\mathfrak{p}}$ is a compact neighborhood of the identity of $K_{\mathfrak{p}}^{\times}$ for the finite primes $\mathfrak{p}$, say $W_{\mathfrak{p}} = U_{\mathfrak{p}}$, then the direct Tychonoff product $\prod_{\mathfrak{p}} W_{\mathfrak{p}}$ is a compact neighborhood of the identity of I_K. This shows that I_K is locally compact.

(7.2) Proposition. $K^{\times}$ *is a discrete and therefore closed subgroup of I_K.*

Proof. It obviously suffices to show that the identity $1 \in I_K$ has an open neighborhood which aside from 1 contains no other principal idèle. Such a neighborhood is given by the set

$$\mathfrak{U} = \{\mathfrak{a} \in I_K \mid |\mathfrak{a}_{\mathfrak{p}} - 1|_{\mathfrak{p}} < 1 \text{ for } \mathfrak{p} \in S, |\mathfrak{a}_{\mathfrak{p}}|_{\mathfrak{p}} = 1 \text{ for } \mathfrak{p} \notin S\},$$

where S denotes a finite set of primes containing all the infinite primes. If there were a principal idèle $x \in \mathfrak{U}$, $x \neq 1$, then we would have

$$\prod_{\mathfrak{p}} |x - 1|_{\mathfrak{p}} = \prod_{\mathfrak{p} \in S} |x - 1|_{\mathfrak{p}} \cdot \prod_{\mathfrak{p} \notin S} |x - 1|_{\mathfrak{p}} < \prod_{\mathfrak{p} \notin S} |x - 1|_{\mathfrak{p}} \leq \prod_{\mathfrak{p} \notin S} \max\{|x|_{\mathfrak{p}}, 1\} = 1,$$

which is a contradiction to the product formula.

Since $K^{\times}$ is closed in I_K, the quotient $C_K = I_K/K^{\times}$ is also a Hausdorff topological group; of course, since I_K is locally compact, so is C_K. The canonical homomorphism $I_K \to C_K$ is continuous and takes open sets to open sets.

For each prime $\mathfrak{p}$, we consider the homomorphism

$$\mathfrak{n}_{\mathfrak{p}} : K_{\mathfrak{p}}^{\times} \longrightarrow I_K \, ,$$

that maps $x \in K_{\mathfrak{p}}^{\times}$ to the idèle $\mathfrak{n}_{\mathfrak{p}}(x) \in I_K$ which at the prime $\mathfrak{p}$ has the component $x \in K_{\mathfrak{p}}^{\times}$, and at all other primes has component 1. If we further map $x \in K_{\mathfrak{p}}^{\times}$ to the idèle class $\overline{\mathfrak{n}}_{\mathfrak{p}}(x) = \mathfrak{n}_{\mathfrak{p}}(x) \cdot K^{\times} \in C_K$ represented by $\mathfrak{n}_{\mathfrak{p}}(x)$, then we obtain a homomorphism

$$\overline{\mathfrak{n}}_{\mathfrak{p}} : K_{\mathfrak{p}}^{\times} \longrightarrow C_K \, ,$$

for which we have the following result:

(7.3) Proposition. *The homomorphism*

$$\overline{\mathfrak{n}}_{\mathfrak{p}} : K_{\mathfrak{p}}^{\times} \longrightarrow C_K$$

is a topological embedding of $K_{\mathfrak{p}}^{\times}$ into C_K.

In fact, from $\overline{\mathfrak{n}}_{\mathfrak{p}}(x) = 1$, i.e., from $\mathfrak{n}_{\mathfrak{p}}(x) \in K^{\times}$, it follows immediately that $x = 1$, so that we are dealing with an injection which is trivially topological.

We associate with each idèle $\mathfrak{a} \in I_K$ its **absolute value**

$$|\mathfrak{a}| = \prod_{\mathfrak{p}} |\mathfrak{a}_{\mathfrak{p}}|_{\mathfrak{p}} \in \mathbb{R}_+^{\times},$$

and obtain a (clearly continuous) homomorphism of I_K to the group $\mathbb{R}_+^{\times}$ of positive real numbers. Of course this homomorphism is surjective; indeed, the group $\mathbb{R}_+^{\times}$ is already exhausted by the absolute value of the idèles of the form $\mathfrak{n}_{\mathfrak{p}}(\mathfrak{a}_{\mathfrak{p}}) = (\ldots, 1, 1, 1, \mathfrak{a}_{\mathfrak{p}}, 1, 1, 1, \ldots)$, $\mathfrak{a}_{\mathfrak{p}} \in K_{\mathfrak{p}}^{\times}$, where $\mathfrak{p}$ is an infinite prime.

We denote by I_K^0 the (closed) kernel of this homomorphism, i.e., the group of idèles of absolute value 1. By the product formula (cf. §1, p. 113) it contains the principal idèle group $K^{\times}$. Thus the absolute value yields a (continuous) homomorphism from the idèle class group C_K onto $\mathbb{R}_+^{\times}$ with (closed) kernel $C_K^0 = I_K^0/K^{\times}$. The group C_K^0 plays a very similar role in C_K as the unit group $U_{\mathfrak{p}} = \{x \in K_{\mathfrak{p}}^{\times} \mid |x|_{\mathfrak{p}} = 1\}$ in the multiplicative group of a local field $K_{\mathfrak{p}}$. The decomposition $K_{\mathfrak{p}}^{\times} \cong U_{\mathfrak{p}} \times \mathbb{Z}$ corresponds to the following

(7.4) Proposition. $C_K = C_K^0 \times \Gamma_K$ with $\Gamma_K \cong \mathbb{R}_+^{\times}$.

Proof. We have to find a splitting for the group extension

$$1 \longrightarrow C_K^0 \longrightarrow C_K \overset{|\,|}{\longrightarrow} \mathbb{R}_+^{\times} \longrightarrow 1,$$

i.e., we have to find an injection $\mathbb{R}_+^{\times} \to C_K$ which when composed with the absolute value map $|\,| : C_K \to \mathbb{R}_+^{\times}$ yields the identity on $\mathbb{R}_+^{\times}$. To do this, we choose an infinite prime $\mathfrak{p}$ and consider the embedding

$$\overline{\mathfrak{n}}_{\mathfrak{p}} : K_{\mathfrak{p}}^{\times} \longrightarrow C_K.$$

Now $K_{\mathfrak{p}}^{\times}$ contains $\mathbb{R}_+^{\times}$ as a subgroup. If $\mathfrak{p}$ is real, then $\overline{\mathfrak{n}}_{\mathfrak{p}} : \mathbb{R}_+^{\times} \to C_K$ is an injection of the desired form, since $|\mathfrak{n}_{\mathfrak{p}}(x)| = |x|_{\mathfrak{p}} = x \in \mathbb{R}_+^{\times}$. If $\mathfrak{p}$ is complex, then $|\mathfrak{n}_{\mathfrak{p}}(x)| = |x|_{\mathfrak{p}} = x^2$, and one has to choose the map $x \mapsto \overline{\mathfrak{n}}_{\mathfrak{p}}(\sqrt[2]{x})$.

Let us mention that there is no distinguished subgroup of representatives in C_K for the factor group $C_K/C_K^0 \cong \mathbb{R}_+^{\times}$.

We will now show that, similar to the the unit group in the local case, the group C_K^0 is compact. For this we need the following lemma.

(7.5) Lemma. *For every prime $\mathfrak{p}$ of the field K let an $\alpha_\mathfrak{p} \in |K_\mathfrak{p}^\times|_\mathfrak{p}$ (the value group of $|\ |_\mathfrak{p}$) be given such that*

1) $\alpha_\mathfrak{p} = 1$ for almost all $\mathfrak{p}$,
2) $\prod_\mathfrak{p} \alpha_\mathfrak{p} \geq \sqrt{|\Delta|}$, where Δ is the discriminant of K.

Then there exists an $x \in K^\times$ with $|x|_\mathfrak{p} \leq \alpha_\mathfrak{p}$ for all $\mathfrak{p}$.

Proof. We set $\alpha_\mathfrak{p} = |\pi_\mathfrak{p}^{e_\mathfrak{p}}|_\mathfrak{p}$ for $\mathfrak{p} \nmid \infty$, where $\pi_\mathfrak{p} \in K_\mathfrak{p}$ is a prime element for $\mathfrak{p}$. Because of 1) $e_\mathfrak{p} = 0$ for almost all $\mathfrak{p}$, and we can consider the ideal $\mathfrak{A} = \prod_{\mathfrak{p} \nmid \infty} \mathfrak{p}^{e_\mathfrak{p}}$, which, because $\alpha_\mathfrak{p} = \mathfrak{N}(\mathfrak{p}^{e_\mathfrak{p}})^{-1}$, has absolute norm

$$\mathfrak{N}(\mathfrak{A}) = \left(\prod_{\mathfrak{p} \nmid \infty} \alpha_\mathfrak{p}\right)^{-1}.$$

Let $a_1, \ldots, a_n$ be an integral basis of $\mathfrak{A}, \gamma_1, \ldots, \gamma_n$ the embeddings of K into the field $\mathbb{C}$ of complex numbers, and $\mathfrak{p}_i$ the infinite prime of K attached to γ_i. We consider the linear forms

$$L_i(x_1, \ldots, x_n) = \sum_{k=1}^{n} x_k \cdot \gamma_i(a_k), \quad i = 1, \ldots, n.$$

If γ_i is real, we set $L_i' = L_i$, $\beta_i = \alpha_{\mathfrak{p}_i}$; in the other case when γ_i and γ_j, $i < j$, are complex conjugates, we set $L_i = L_i' + \sqrt{-1} \cdot L_j'$, thus $L_j = \bar{L}_i = L_i' - \sqrt{-1} \cdot L_j'$, and $\beta_i = \beta_j = \sqrt{\alpha_{\mathfrak{p}_i}/2}$. If r is the number of complex primes, it follows that

$$|\det(L_1', \ldots, L_n')| = \left|\left(\frac{1}{-2\sqrt{-1}}\right)^r \det(L_1, \ldots, L_n)\right| = \frac{1}{2^r} \mathfrak{N}(\mathfrak{A}) \cdot \sqrt{|\Delta|}$$

$$= \frac{1}{2^r}\left(\prod_{\mathfrak{p} \nmid \infty} \alpha_\mathfrak{p}\right)^{-1} \cdot \sqrt{|\Delta|} \leq \frac{1}{2^r} \prod_{\mathfrak{p} | \infty} \alpha_\mathfrak{p} = \prod_{i=1}^{n} \beta_i.$$

The well-known Minkowski Theorem about Linear Forms (cf. e.g. [22], §17) now yields an integral vector $\mathfrak{Z} = (m_1, \ldots, m_n) \in \mathbb{Z}^n$, such that $|L_i'(\mathfrak{Z})| \leq \beta_i$, $i = 1, \ldots, n$, hence $|L_i(\mathfrak{Z})| \leq \alpha_{\mathfrak{p}_i}$ (resp. $\leq \sqrt{\alpha_{\mathfrak{p}_i}}$) if $\mathfrak{p}_i$ is real (resp. complex). Set $x = m_1 a_1 + \cdots + m_n a_n \in \mathfrak{A}$. Then $|x|_\mathfrak{p} \leq \alpha_\mathfrak{p}$ for the finite primes by construction of $\mathfrak{A}$, and for the infinite primes because $|\gamma_i(x)| = |L_i(\mathfrak{Z})|$. $\square$

(7.6) Theorem. *The group C_K^0 is compact.*

Proof. To simplify notation we set $I = I_K$, $I^0 = I_K^0$, $C = C_K$, and $C^0 = C_K^0$. We consider the set

$$\mathcal{K} = \prod_{\text{all } \mathfrak{p}} \mathcal{K}_\mathfrak{p}, \quad \text{with } \mathcal{K}_\mathfrak{p} := \{a \in K_\mathfrak{p}^\times \mid 1/\sqrt{|\Delta|} \leq |a|_\mathfrak{p} \leq \sqrt{|\Delta|}\},$$

which is compact as the direct Tychonoff product of compact spaces. We have $\mathcal{K}_\mathfrak{p} = U_\mathfrak{p}$ for all non-archimedian $\mathfrak{p}$ with $N(\mathfrak{p}) > \sqrt{|\Delta|}$, and since these are all but finitely many, $\mathcal{K}$ is a subset of I. Since I^0 is closed in I, $\mathcal{K}^0 := \mathcal{K} \cap I^0$

is thus a compact subset of I^0. To prove C^0 is compact it now suffices to show that $\mathcal{K}^0$ is mapped by the (continuous) transformation $I^0 \to C^0$ onto the entire group $C^0 = I^0/K^\times$, i.e., that for each idèle $\mathfrak{a} \in I^0$ there exists an $x \in K^\times$ with $x\mathfrak{a}^{-1} \in \mathcal{K}^0$. For this we choose a fixed infinite prime $\mathfrak{q}$ and set

$$\alpha_\mathfrak{q} = \sqrt{|\Delta|} \cdot |\mathfrak{a}_\mathfrak{q}|_\mathfrak{q} \quad \text{and} \quad \alpha_\mathfrak{p} = |\mathfrak{a}_\mathfrak{p}|_\mathfrak{p} \quad \text{for all } \mathfrak{p} \neq \mathfrak{q}.$$

Because $\mathfrak{a} \in I^0$, $\prod_\mathfrak{p} \alpha_\mathfrak{p} = \sqrt{|\Delta|}$, and by (7.5) there is an $x \in K^\times$ with $|x|_\mathfrak{p} \leq \alpha_\mathfrak{p}$, so that $|x \cdot \mathfrak{a}_\mathfrak{p}^{-1}|_\mathfrak{p} \leq 1 \ (\leq \sqrt{|\Delta|})$, $\mathfrak{p} \neq \mathfrak{q}$, resp. $|x \cdot \mathfrak{a}_\mathfrak{q}^{-1}|_\mathfrak{q} \leq \sqrt{|\Delta|}$. By the product formula we obtain further

$$1 = \prod_\mathfrak{p} |x \cdot \mathfrak{a}_\mathfrak{p}^{-1}|_\mathfrak{p} = |x \cdot \mathfrak{a}_\mathfrak{q}^{-1}|_\mathfrak{q} \cdot \prod_{\mathfrak{p} \neq \mathfrak{q}} |x \cdot \mathfrak{a}_\mathfrak{p}^{-1}|_\mathfrak{p},$$

therefore $|x \cdot \mathfrak{a}_\mathfrak{q}^{-1}|_\mathfrak{q} = (\prod_{\mathfrak{p} \neq \mathfrak{q}} |x \cdot \mathfrak{a}_\mathfrak{p}^{-1}|_\mathfrak{p})^{-1} \geq 1 \geq 1/\sqrt{|\Delta|}$; for $\mathfrak{p} \neq \mathfrak{q}$ it follows that $|x \cdot \mathfrak{a}_\mathfrak{p}^{-1}|_\mathfrak{p} \geq \prod_{\mathfrak{p}' \neq \mathfrak{q}} |x \cdot \mathfrak{a}_{\mathfrak{p}'}^{-1}|_{\mathfrak{p}'} = 1/|x \cdot \mathfrak{a}_\mathfrak{q}^{-1}|_\mathfrak{q} \geq 1/\sqrt{|\Delta|}$.

Altogether we therefore have

$$1/\sqrt{|\Delta|} \leq |x \cdot \mathfrak{a}_\mathfrak{p}^{-1}|_\mathfrak{p} \leq \sqrt{|\Delta|} \quad \text{for all } \mathfrak{p},$$

i.e., $x \cdot \mathfrak{a}^{-1} \in \mathcal{K}^0$. Thus all is proved.[*]

If S is a finite set of primes of K, then we let U_K^S be the idèle group

$$U_K^S = \{\mathfrak{a} \in I_K | \mathfrak{a}_\mathfrak{p} = 1 \text{ for } \mathfrak{p} \in S; \mathfrak{a}_\mathfrak{p} \in U_\mathfrak{p} \text{ for } \mathfrak{p} \notin S\} \subseteq I_K^S$$

and

$$\overline{U}_K^S = U_K^S \cdot K^\times / K^\times \subseteq C_K.$$

In each neighborhood of the identity of C_K there is a group $\overline{U}_K^S$ [20]; this can be seen from the following observations:

- The groups

$$\prod_{\mathfrak{p} \in S} W_\mathfrak{p} \times \prod_{\mathfrak{p} \notin S} U_\mathfrak{p} \subseteq I_K$$

  form a fundamental system of neighborhoods of the identity of I_K, where $W_\mathfrak{p}$ is a fundamental system of neighborhoods of the identity of $K_\mathfrak{p}^\times$ and S runs through all the finite primes (cf. p. 156),

- they always contain one of the groups U_K^S,

- passing from I_K to C_K, a fundamental system of neighborhoods is transformed into a fundamental system of neighborhoods, and U_K^S into $\overline{U}_K^S$.

---

[*] Remark of the editor: The proof does not use the finiteness of the ideal class group. But this finiteness follows from (7.6): The composition of the homomorphisms $C_K^0 \hookrightarrow C_K \twoheadrightarrow J_K/P_K$ is surjective and continuous with respect to the discrete topology on J_K/P_K. Consequently J_K/P_K is discrete as well as compact, and thus finite.

[20] Note that the $\overline{U}_K^S$ themselves are not open. But since the groups $U_\mathfrak{p}$ are closed in $K_\mathfrak{p}^\times$, the $\overline{U}_K^S$ are closed.

The following theorem on Kummer extensions is the key ingredient for the proof of the fundamental Existence Theorem announced above.

(7.7) Theorem. *Let K be a field which contains the n-th roots of unity. If S is a finite set of primes of K such that*

1) *S contains all the infinite primes and all the primes lying above the prime numbers dividing n,*

2) *$I_K = I_K^S \cdot K^\times$,*

then $C_K^n \cdot \overline{U}_K^S$ is the norm group of the Kummer extension $T = K(\sqrt[n]{K^S})|K$.

Addendum. *If K does not contain the n-th roots of unity, $C_K^n \cdot \overline{U}_K^S$ is still a norm group.*

Proof. By (1.4) $\chi(G_{T|K}) \cong K^S \cdot (K^\times)^n/(K^\times)^n \cong K^S/(K^S)^n$. Since K^S is finitely generated of rank $N - 1 = |S| - 1$ (cf. (1.1)) and contains the n-th roots of unity, $K^S/(K^S)^n$ is the direct product of N cyclic groups of order n. Therefore $G_{T|K}$ is also the direct product of N cyclic groups of order n.

It follows from this that for each $\overline{\mathfrak{a}}^n \in C_K^n$ we have $(\overline{\mathfrak{a}}^n, T|K) = (\overline{\mathfrak{a}}, T|K)^n = 1$, i.e., $\overline{\mathfrak{a}}^n \in N_{T|K}C_T$, so that $C_K^n \subseteq N_{T|K}C_T$. Furthermore, every idèle $\mathfrak{a} \in U_K^S$ is a norm idèle of the extension $T|K$. To see this we must convince ourselves by (3.4) that $\mathfrak{a}_\mathfrak{p}$ is a norm of the local extension $K_\mathfrak{p}(\sqrt[n]{K^S})|K_\mathfrak{p}$ for every $\mathfrak{p}$. For $\mathfrak{p} \in S$ this is trivial because $\mathfrak{a}_\mathfrak{p} = 1$. If $\mathfrak{p} \notin S$, then $\mathfrak{a}_\mathfrak{p} \in U_\mathfrak{p}$, and $\mathfrak{a}_\mathfrak{p}$ is by II, (4.4) a norm if the extension $K_\mathfrak{p}(\sqrt[n]{K^S})|K_\mathfrak{p}$ is unramified. But this is immediately clear, since every $a \in K^S$ is a unit for $\mathfrak{p} \notin S$, and since n is relatively prime to the characteristic char $\overline{K}_\mathfrak{p}$ of the residue field if $\mathfrak{p} \notin S$, the equation $X^n - a = 0$ is separable over the residue field $\overline{K}_\mathfrak{p}$, so that $K_\mathfrak{p}(\sqrt[n]{a})|K_\mathfrak{p}$ is unramified. It follows that $\overline{U}_K^S \subseteq N_{T|K}C_T$ and therefore

$$C_K^n \cdot \overline{U}_K^S \subseteq N_{T|K}C_T.$$

We show this inclusion is an equality using an index argument. We have

$$(C_K : N_{T|K}C_T) = |G_{T|K}| = (K^S : (K^S)^n) = n^N, \quad N = |S|$$

by the Reciprocity law. On the other hand,

$$(*) \quad (C_K : C_K^n \cdot \overline{U}_K^S) = (I_K^S \cdot K^\times : (I_K^S)^n \cdot U_K^S \cdot K^\times)$$
$$= (I_K^S : (I_K^S)^n \cdot U_K^S)/((I_K^S \cap K^\times) : ((I_K^S)^n \cdot U_K^S \cap K^\times))\ ^{21)}.$$

Here the index in the numerator is

$$(I_K^S : (I_K^S)^n \cdot U_K^S) = \prod_{\mathfrak{p} \in S}(K_\mathfrak{p}^\times : (K_\mathfrak{p}^\times)^n),$$

$^{21)}$ For this cf. the footnote $^{12)}$ on p. 133.

since the map $I_K^S \to \prod_{\mathfrak{p} \in S} K_\mathfrak{p}^\times / (K_\mathfrak{p}^\times)^n$, $\mathfrak{a} \mapsto \prod_{\mathfrak{p} \in S} \mathfrak{a}_\mathfrak{p} \cdot (K_\mathfrak{p}^\times)^n$ is obviously surjective, and its kernel consists of those idèles $\mathfrak{a} \in I_K^S$, for which $\mathfrak{a}_\mathfrak{p} \in (K_\mathfrak{p}^\times)^n$ for $\mathfrak{p} \in S$; but these are precisely the idèles in $(I_K^S)^n \cdot U_K^S$.

By II, (3.7) we obtain, using that $|n|_\mathfrak{p} = 1$ for $\mathfrak{p} \notin S$, the identity

$$(I_K^S : (I_K^S)^n \cdot U_K^S) = \prod_{\mathfrak{p} \in S} (K_\mathfrak{p}^\times : (K_\mathfrak{p}^\times)^n) = \prod_{\mathfrak{p} \in S} n^2 \cdot |n|_\mathfrak{p} = n^{2N} \cdot \prod_\mathfrak{p} |n|_\mathfrak{p} = n^{2N}.$$

For the index in the denominator of the equation $(*)$ we have $I_K^S \cap K^\times = K^S$ and $(I_K^S)^n \cdot U_K^S \cap K^\times = (K^S)^n$. The first statement is clear. For the second, it is clear that we have in any case the inclusion $(K^S)^n \subseteq (I_K^S)^n \cdot U_K^S \cap K^\times$.

To show the opposite inclusion, let $x \in (I_K^S)^n \cdot U_K^S \cap K^\times$, thus $x = \mathfrak{a}^n \cdot u$, $\mathfrak{a} \in I_K^S$, $u \in U_K^S$. We form the field $K(\sqrt[n]{x})$ and show $K(\sqrt[n]{x}) = K$. If $\mathfrak{b} \in I_K^S$, then $\mathfrak{b}$ is always a norm-idèle of $K(\sqrt[n]{x})|K$. In fact, if $\mathfrak{p} \in S$, then $\mathfrak{b}_\mathfrak{p} \in K_\mathfrak{p}^\times$ is a norm because $K_\mathfrak{p}(\sqrt[n]{x}) = K_\mathfrak{p}(\sqrt[n]{\mathfrak{a}_\mathfrak{p}^n}) = K_\mathfrak{p}$. If $\mathfrak{p} \notin S$ the same holds, since $\mathfrak{b}_\mathfrak{p} \in U_\mathfrak{p}$ and $K_\mathfrak{p}(\sqrt[n]{x}) = K_\mathfrak{p}(\sqrt[n]{u_\mathfrak{p}})|K_\mathfrak{p}$ because $\mathfrak{p} \nmid n$ is unramified (cf. II, (4.4)). If we keep in mind that $I_K = I_K^S \cdot K^\times$, we have thus shown that

$$N_{K(\sqrt[n]{x})|K} C_{K(\sqrt[n]{x})} = C_K,$$

which implies by the Reciprocity Law that $K = K(\sqrt[n]{x})$. Hence we have $\sqrt[n]{x} = y \in K^\times$, $x = y^n \in (K^\times)^n \cap K^S = (K^S)^n$.

From the equation $(*)$ we now obtain

$$(C_K : C_K^n \cdot \overline{U}_K^S) = n^{2N} / (K^S : (K^S)^n) = n^{2N} / n^N = n^N = (C_K : N_{T|K} C_T),$$

and therefore $N_{T|K} C_T = C_K^n \cdot \overline{U}_K^S$.

If we drop the assumption on the roots of unity, then $C_K^n \cdot \overline{U}_K^S$ nevertheless turns out to be a norm group, as claimed in the addendum. In fact, if the field K does not contain the n-th roots of unity, then we adjoin these and let $K'|K$ be the resulting extension. If S' is a finite set of primes of K' which contain all the primes above the primes in S and in addition is sufficiently large so that $I_{K'} = I_{K'}^{S'} \cdot K'^\times$, then by what we just proved $C_{K'}^n \cdot \overline{U}_{K'}^{S'}$ is the norm group of a normal extension $L'|K'$. If L is the smallest normal extension of K containing L', then we have

$$N_{L|K} C_L = N_{K'|K}(N_{L'|K'}(N_{L|L'} C_L)) \subseteq N_{K'|K}(N_{L'|K'} C_{L'}) =$$

$$N_{K'|K}(C_{K'}^n \cdot \overline{U}_{K'}^{S'}) = (N_{K'|K} C_{K'})^n \cdot N_{K'|K} \overline{U}_{K'}^{S'} \subseteq C_K^n \cdot \overline{U}_K^S.$$

Thus $C_K^n \cdot \overline{U}_K^S$, as a group which contains the norm group $N_{L|K} C_L$, is itself a norm group (cf. (6.14)).

(7.8) Existence Theorem. *The norm groups of C_K are precisely the closed subgroups of finite index.*

Proof. Let $\mathcal{N}_L = N_{L|K}C_L \subseteq C_K$ be the norm group of a normal extension $L|K$. By the Reciprocity Law the index $(C_K : \mathcal{N}_L) = |G_{L|K}^{ab}|$ is finite. The norm map $N_{L|K} : C_L \to C_K$ is clearly continuous. We have $C_K = C_K^0 \times \Gamma_K$, $C_L = C_L^0 \times \Gamma_L$, with $\Gamma_K, \Gamma_L \cong \mathbb{R}_+^\times$. The injection $\mathbb{R}_+^\times \to C_K$ from the proof of (7.4) obviously also yields a group of representatives for C_L/C_L^0; therefore we can assume that $\Gamma_K = \Gamma_L \subseteq C_L$. It follows that

$$N_{L|K}C_L = N_{L|K}C_L^0 \times N_{L|K}\Gamma_K = N_{L|K}C_L^0 \times \Gamma_K^n = N_{L|K}C_L^0 \times \Gamma_K.$$

The image of the compact group C_L^0 in C_K is compact, therefore closed, and since $\Gamma_K \subseteq C_K$ is also closed, $N_{L|K}C_L$ is in fact closed.

Conversely, let $\mathcal{N} \subseteq C_K$ be a closed subgroup of finite index $(C_K : \mathcal{N}) = n$. Then in any case $C_K^n \subseteq \mathcal{N}$.

Moreover, $\mathcal{N}$ is (as the complement of its finitely many closed cosets) also open, and therefore contains one of the groups $\bar{U}_K^S$ (cf. the remark on p. 160). But $C_K^n \cdot \bar{U}_K^S$ is by the addendum to (7.7) a norm group for sufficiently large S, and by (6.14) the same holds for the group $\mathcal{N}$ as it contains a norm group.

We have proved using (7.7) that for every closed subgroup $\mathcal{N} \subseteq C_K$ of finite index there exists a normal extension $L|K$ with norm group $N_{L|K}C_L = \mathcal{N}$, which is precisely the fundamental existence statement.

From the identification of the norm groups given by the Existence Theorem we obtain without difficulty a further characterization of these groups which is of a predominantly arithmetic nature. It is the idèle theoretic version of the formulation of the Existence Theorem in the classical theory using ideals. [22].

By a **modulus** $\mathfrak{m}$ we mean a formal product

$$\mathfrak{m} = \prod_{\mathfrak{p}} \mathfrak{p}^{n_\mathfrak{p}}$$

of prime powers, such that $n_\mathfrak{p} \geq 0$ and $n_\mathfrak{p} = 0$ for almost all primes $\mathfrak{p}$; for the infinite primes $\mathfrak{p}$ we allow only the exponents $n_\mathfrak{p} = 0$ and 1.

For a prime $\mathfrak{p}$ of K let

$$U_\mathfrak{p}^{n_\mathfrak{p}} = \begin{cases} \text{the } n_\mathfrak{p}\text{-th unit group of } K_\mathfrak{p}, \ U_\mathfrak{p}^0 = U_\mathfrak{p}, \text{ when } \mathfrak{p} \nmid \infty, \\ \mathbb{R}_+^\times \subseteq K_\mathfrak{p}^\times, \text{ when } \mathfrak{p} \text{ is real and } n_\mathfrak{p} = 1, \\ \mathbb{R}^\times = K_\mathfrak{p}^\times, \text{ when } \mathfrak{p} \text{ is real and } n_\mathfrak{p} = 0, \\ \mathbb{C}^\times = K_\mathfrak{p}^\times, \text{ when } \mathfrak{p} \text{ is complex.} \end{cases}$$

---

[22] Cf. [17], Teil I, §4, Satz 1.

If $\mathfrak{a}_\mathfrak{p} \in K_\mathfrak{p}^\times$, then let

$$\mathfrak{a}_\mathfrak{p} \equiv 1 \bmod \mathfrak{p}^{n_\mathfrak{p}} \Longleftrightarrow \mathfrak{a}_\mathfrak{p} \in U_\mathfrak{p}^{n_\mathfrak{p}}\ .$$

For a finite prime $\mathfrak{p}$ and $n_\mathfrak{p} \geq 1$ (resp. $n_\mathfrak{p} = 0$) this means ordinary congruence (resp. $\mathfrak{a}_\mathfrak{p} \in U_\mathfrak{p}$), for a real prime $\mathfrak{p}$ with exponent $n_\mathfrak{p} = 1$ this is the positivity condition $a_\mathfrak{p} > 0$, and for the remaining cases where $\mathfrak{p}$ is real and $n_\mathfrak{p} = 0$ or $\mathfrak{p}$ is complex, there is no restriction.

If $\mathfrak{m} = \prod_\mathfrak{p} \mathfrak{p}^{n_\mathfrak{p}}$ is a modulus, then for an idèle $\mathfrak{a} \in I_K$ we set

$$\mathfrak{a} \equiv 1 \bmod \mathfrak{m} \Longleftrightarrow \mathfrak{a}_\mathfrak{p} \equiv 1 \bmod \mathfrak{p}^{n_\mathfrak{p}} \text{ for all } \mathfrak{p},$$

and consider the group

$$I_K^\mathfrak{m} = \{\mathfrak{a} \in I_K \mid \mathfrak{a} \equiv 1 \bmod \mathfrak{m}\} = \prod_\mathfrak{p} U_\mathfrak{p}^{n_\mathfrak{p}} \subseteq I_K.$$

If in particular $\mathfrak{m} = 1$, then obviously

$$I_K^1 = I_K^{S_\infty} = \prod_{\mathfrak{p}\mid\infty} K_\mathfrak{p}^\times \times \prod_{\mathfrak{p}\nmid\infty} U_\mathfrak{p},$$

where S_∞ is the set of infinite primes of K.

We call the idèle class group

$$C_K^\mathfrak{m} = I_K^\mathfrak{m} \cdot K^\times / K^\times \subseteq C_K$$

the **congruence subgroup mod $\mathfrak{m}$** of C_K. The factor group $C_K/C_K^\mathfrak{m}$ is also called the **ray class group mod $\mathfrak{m}$**. If particular, if $\mathfrak{m} = 1$, then we have

$$C_K/C_K^1 = I_K/K^\times \big/ I_K^1 \cdot K^\times / K^\times \cong I_K/I_K^{S_\infty} \cdot K^\times.$$

Hence by (2.3) the ray class group mod 1 is isomorphic to the ideal class group J_K/P_K, and its order is equal to the class number h of K.

(7.9) Theorem. *The norm groups of C_K are precisely the groups containing the congruence subgroups $C_K^\mathfrak{m}$.*

Proof. The index $(C_K : C_K^\mathfrak{m}) = (C_K : C_K^1) \cdot (C_K^1 : C_K^\mathfrak{m}) = h \cdot (C_K^1 : C_K^\mathfrak{m}) = h \cdot (I_K^1 \cdot K^\times : I_K^\mathfrak{m} \cdot K^\times) \leq h \cdot (I_K^1 : I_K^\mathfrak{m}) = h \cdot \prod_\mathfrak{p}(U_\mathfrak{p} : U_\mathfrak{p}^{n_\mathfrak{p}})$ is finite. Since $I_K^\mathfrak{m} = \prod_\mathfrak{p} U_\mathfrak{p}^{n_\mathfrak{p}}$ is open in I_K, the image $C_K^\mathfrak{m}$ is also open and therefore closed in C_K. Thus by (7.8) the congruence subgroups, and all groups containing them, are norm groups.

Assume conversely that $\mathcal{N}$ is a norm group of C_K, so that by (7.8) $\mathcal{N}$ is a closed subgroup of finite index. Then $\mathcal{N}$ is also open and has an open preimage $\mathcal{I}$ in I_K. This open preimage $\mathcal{I}$ contains an open subgroup W of the form

$$W = \prod_{\mathfrak{p}\in S} W_\mathfrak{p} \times \prod_{\mathfrak{p}\notin S} U_\mathfrak{p},$$

where S is a finite set of primes S and each $W_{\mathfrak{p}}$ is an open neighborhood of the identity of $K_{\mathfrak{p}}^{\times}$; in fact, these groups form a fundamental system of neighborhoods of the identity of I_K (cf. p. 156). If $\mathfrak{p}$ is finite, then we can assume $W_{\mathfrak{p}} = U_{\mathfrak{p}}^{n_{\mathfrak{p}}}$, since the unit groups $U_{\mathfrak{p}}^{n_{\mathfrak{p}}} \subseteq K_{\mathfrak{p}}^{\times}$ form a basis. If $\mathfrak{p}$ is infinite, then the open sets $W_{\mathfrak{p}}$ generate the entire group $K_{\mathfrak{p}}^{\times}$ or the group $\mathbb{R}_{+}^{\times}$ in the real case. The group generated by W is thus a group $I_K^{\mathfrak{m}} \subseteq I$ for an appropriate modulus $\mathfrak{m}$, and $\mathcal{N}$ is a group containing $C_K^{\mathfrak{m}} = I_K^{\mathfrak{m}} \cdot K^{\times}/K^{\times}$.

The (abelian) class field $L|K$ associated with the norm group $C_K^{\mathfrak{m}}$ is called the **ray class field mod $\mathfrak{m}$**. By the Reciprocity Law its Galois group $G_{L|K}$ is isomorphic to the ray class group $C_K/C_K^{\mathfrak{m}}$. In case $K = \mathbb{Q}$, we have:

(7.10) Theorem. *Let m be a positive integer, p_{∞} the infinite prime of $\mathbb{Q}$, and $\mathfrak{m}$ the modulus $\mathfrak{m} = m \cdot p_{\infty}$. The ray class field mod $\mathfrak{m}$ is precisely the field $\mathbb{Q}(\zeta_m)$ of m-th roots of unity.*

We point out that this theorem implies that the Existence Theorem for the field $\mathbb{Q}$ is precisely the famous **Theorem of Kronecker:**

(7.11) Theorem. *Every abelian extension of the field $\mathbb{Q}$ of rational numbers is a cyclotomic field, i.e., a subfield of a field of roots of unity $\mathbb{Q}(\zeta)$.*

Proof of (7.10). Let ζ be a primitive m-th root of unity and $m = \prod_p p^{n_p}$. Then $I_{\mathbb{Q}}^{\mathfrak{m}} = \prod_{p \neq p_{\infty}} U_p^{n_p} \times \mathbb{R}_{+}^{\times}$. The group $U_p^{n_p} \subseteq \mathbb{Q}_p^{\times}$ consists of norms in the extension $\mathbb{Q}_p(\zeta)|\mathbb{Q}_p$. In fact, the field $\mathbb{Q}_p(\zeta)$ is the compositum of the field of p^{n_p}-th roots of unity $\mathbb{Q}_p(\zeta_{p^{n_p}})$ and the unramified extension over $\mathbb{Q}_p$ of the m_p'-th roots of unity $\mathbb{Q}_p(\zeta_{m_p'})$ $(m_p' = m/p^{n_p})$. The first field contains the elements in $U_p^{n_p}$ as norms because of II, (7.15), the second because of II, (4.4).

Since the norm group of the compositum $\mathbb{Q}_p(\zeta) = \mathbb{Q}_p(\zeta_{p^{n_p}})\mathbb{Q}_p(\zeta_{m_p'})$ is the intersection of the two norm groups of $\mathbb{Q}_p(\zeta_{p^{n_p}})$ and $\mathbb{Q}_p(\zeta_{m_p'})$, the group $U_p^{n_p}$ is contained in the norm group of $\mathbb{Q}_p(\zeta)|\mathbb{Q}_p$. It follows from (3.4) that $I_{\mathbb{Q}}^{\mathfrak{m}}$ consists of norm idèles of the extension $\mathbb{Q}(\zeta)|\mathbb{Q}$, i.e., $C_{\mathbb{Q}}^{\mathfrak{m}} \subseteq N_{\mathbb{Q}(\zeta)|\mathbb{Q}}C_{\mathbb{Q}(\zeta)}$.

On the other hand,

$$(C_{\mathbb{Q}} : C_{\mathbb{Q}}^{\mathfrak{m}}) = (C_{\mathbb{Q}} : C_{\mathbb{Q}}^{1}) \cdot (C_{\mathbb{Q}}^{1} : C_{\mathbb{Q}}^{\mathfrak{m}}) = (I_{\mathbb{Q}}^{1} \cdot \mathbb{Q}^{\times} : I_{\mathbb{Q}}^{\mathfrak{m}} \cdot \mathbb{Q}^{\times})$$
$$= (I_{\mathbb{Q}}^{1} : I_{\mathbb{Q}}^{\mathfrak{m}})/((I_{\mathbb{Q}}^{1} \cap \mathbb{Q}^{\times}) : (I_{\mathbb{Q}}^{\mathfrak{m}} \cap \mathbb{Q}^{\times})).$$

Since $I_{\mathbb{Q}}^1 = \prod_{p \neq p_\infty} U_p \times \mathbb{R}^\times$ and $I_{\mathbb{Q}}^{\mathfrak{m}} = \prod_{p \neq p_\infty} U_p^{n_p} \times \mathbb{R}_+^\times$, we have $I_{\mathbb{Q}}^1 \cap \mathbb{Q}^\times = \{1, -1\}$ and $I_{\mathbb{Q}}^{\mathfrak{m}} \cap \mathbb{Q}^\times = \{1\}$. We thus obtain for the above index the formula

$$(C_{\mathbb{Q}} : C_{\mathbb{Q}}^{\mathfrak{m}}) = \frac{1}{2} \cdot \prod_{p \neq p_\infty} (U_p : U_p^{n_p}) \cdot (\mathbb{R}^\times : \mathbb{R}_+^\times)$$

$$= \prod_{p \mid m} (U_p : U_p^{n_p}) = \prod_{p \mid m} p^{n_p - 1} \cdot (p - 1) = \varphi(m).$$

But this is the degree $\varphi(m) = [\mathbb{Q}(\zeta) : \mathbb{Q}]$ of the field extension $\mathbb{Q}(\zeta) | \mathbb{Q}$, and it follows that $(C_{\mathbb{Q}} : C_{\mathbb{Q}}^{\mathfrak{m}}) = (C_{\mathbb{Q}} : N_{\mathbb{Q}(\zeta)|\mathbb{Q}} C_{\mathbb{Q}(\zeta)})$, i.e., the congruence subgroup $C_{\mathbb{Q}}^{\mathfrak{m}}$ is in fact the norm group of the cyclotomic field $\mathbb{Q}(\zeta)$.

Because of (7.10) we can think of the ray class fields over a number field K as the fields corresponding to the fields of roots of unity in case $K = \mathbb{Q}$. In this context Kronecker's Theorem (7.11) corresponds to the deep generalization that every abelian extension of K is a subfield of a ray class field. In this sense class field theory appears as a generalization of the theory of cyclotomic fields; in fact, the historical development of class field theory has been guided to a large extend by the example of cyclotomic fields.

With the introduction of ray class fields we obtain a good overview over the lattice of all abelian extensions of a base field K. The ray class fields themselves correspond to the different moduli $\mathfrak{m}$ of K, where the larger modulus corresponds to the smaller congruence subgroup, thus the larger ray class field. More precisely, if $\mathfrak{m}$ and $\mathfrak{m}'$ are two moduli of K, then $\mathfrak{m} \mid \mathfrak{m}'$ implies that the ray class field mod $\mathfrak{m}$ is contained in the ray class field mod $\mathfrak{m}'$. Among all ray class fields over K, there is one which plays a special role but only appears when we get away from the base field $K = \mathbb{Q}$. This is the ray class field mod 1, i.e., the class field $L | K$ associated with the congruence group C_K^1 with modulus $\mathfrak{m} = 1$. It is called the **Hilbert** or also the **absolute class field over K**. Its Galois group is canonically isomorphic to C_K / C_K^1, and therefore to the ideal class group J_K / P_K (cf. p. 164). Its degree $[L : K]$ is equal to the ideal class number h of K. We will discuss the Hilbert class field in the next section.

We use the compactness of the group C_K^0 and the fact that the groups $C_K^n \cdot \overline{U}_K^S$ are norm groups (cf. (7.7) and Addendum) to prove the following theorem about the universal norm residue symbol $(\ , K)$.

(7.12) Theorem. *The universal norm residue symbol*

$$C_K \xrightarrow{(\ ,K)} G_K^{ab}$$

from C_K to the topological Galois group G_K^{ab} of the maximal abelian extension $A_K|K$ is continuous, surjective homomorphism, and its kernel $D_K = \bigcap_L N_{L|K}C_L$ is the group of all infinitely divisible elements of C_K [23]*, i.e.,*

$$D_K = \bigcap_{n=1}^{\infty} C_K^n.$$

Proof. We first prove the last statement. If $\bar{a} \in \bigcap_{n=1}^{\infty} C_K^n$, and if $\mathcal{N}_L = N_{L|K}C_L$ is any norm group, then $\bar{a} = \bar{b}^n \in \mathcal{N}_L$ if $(C_K : \mathcal{N}_L) = n$. Therefore $\bigcap_{n=1}^{\infty} C_K^n \subseteq D_K = \bigcap_L N_{L|K}C_L$.

On the other hand, the groups $C_K^n \cdot \bar{U}_K^S$ are norm groups for sufficiently large S, i.e., $D_K \subseteq C_K^n \cdot \bar{U}_K^S$. For the inclusion $D_K \subseteq \bigcap_{n=1}^{\infty} C_K^n$ it thus suffices to show that $\bigcap_S C_K^n \cdot \bar{U}_K^S = C_K^n$. Let $\bar{a} \in \bigcap_S C_K^n \cdot \bar{U}_K^S$; then for every S we have the representation $\bar{a} = \bar{b}_S^n \cdot \bar{u}_S$, $\bar{b}_S \in C_K$, $\bar{u}_S \in \bar{U}_K^S$. Because $\bigcap_S U_K^S = 1$ we also have $\bigcap_S \bar{U}_K^S = 1$, and this means that the sequence $\bar{u}_S$ for increasing S converges to 1, i.e., the sequence $\bar{a} \cdot \bar{u}_S^{-1} \in C_K^n$ converges to $\bar{a}$: $\bar{a} = \lim_S \bar{a} \cdot \bar{u}_S^{-1}$.

Consider now $C_K = C_K^0 \times \Gamma_K$ ($\Gamma_K \cong \mathbb{R}_+^\times$). Since $C_K \xrightarrow{n} C_K^n$ is continuous, C_K^0 is compact and Γ_K is closed, thus we see that the group

$$C_K^n = (C_K^0)^n \times \Gamma_K^n = (C_K^0)^n \times \Gamma_K$$

is closed, so that $\bar{a} = \lim_S \bar{a} \cdot \bar{u}_S^{-1} \in C_K^n$. Hence $D_K \subseteq \bigcap_S C_K^n \cdot \bar{U}_K^S = C_K^n$ for all n, which shows that $D_K \subseteq \bigcap_{n=1}^{\infty} C_K^n$, and proves $D_K = \bigcap_{n=1}^{\infty} C_K^n$.

In particular, since $C_K = C_K^0 \times \Gamma_K$ and $\Gamma_K \cong \mathbb{R}_+^\times$ consists of infinitely divisible elements, this group is contained in the kernel D_K of $(\ ,K)$, i.e.,

$$(C_K, K) = (C_K^0, K) \subseteq G_K^{ab}.$$

Now if the homomorphism $C_K \xrightarrow{(\ ,K)} G_K^{ab}$ is continuous, then the compactness of C_K^0 implies that the image $(C_K^0, K) = (C_K, K)$ is closed in G_K^{ab}, and because of denseness we have $(C_K, K) = G_K^{ab}$, which will prove surjectivity.

But to show the map $(\ ,K)$ is continuous is almost trivial. If H is an open subgroup of G_K^{ab}, thus a closed subgroup of finite index, and L is the fixed field of H, then the norm group $\mathcal{N}_L = N_{L|K}C_L \subseteq C_K$ is open, and because $(\mathcal{N}_L, L|K) = (\mathcal{N}_L, K) \cdot H = 1$ it is mapped by $(\ ,K)$ into H.

In addition to the above characterization, the group D_K has a purely topological description as the connected component of the identity of C_K. For the proof of this non-trivial fact we refer the interested reader to [2], Ch. 9.

---

[23] An element $\bar{a} \in C_K$ is infinitely divisible, if for every natural number n there exists a $\bar{b} \in C_K$ with $\bar{a} = \bar{b}^n$.

§ 8. The Decomposition Law

Many of the deepest statements of number theory find their common expression in Artin's Reciprocity Law. For example, and without giving details, one can regard Gauss's Reciprocity Law for quadratic residues as a special case[24], and more generally, the theory of higher power residues is dominated by Artin's Reciprocity Law. Another important application concerns the question of which ideals of a base field K become principal in an extension field L, which we will return to later. The most important consequence, however, is the answer to the question of how the prime ideals $\mathfrak{p}$ of a basis field K split in an abelian extension. For this, we consider instead of the prime ideal $\mathfrak{p}$ of K an associated "prime idèle" by choosing a prime element $\pi \in K_{\mathfrak{p}}$ and forming the idèle $\mathfrak{n}_{\mathfrak{p}}(\pi) = (\dots, 1, 1, \pi, 1, 1, \dots)$. If we first disregard the finitely many ramified primes, then the decomposition of the prime ideal $\mathfrak{p}$ in the abelian extension $L|K$ can be immediately read off from a relation between the prime idèle $\mathfrak{n}_{\mathfrak{p}}(\pi)$ in the norm group $\mathcal{N}_L = N_{L|K} C_K \subseteq C_K$ which determines L, namely, simply from the order of the idèle class $\overline{\mathfrak{n}}_{\mathfrak{p}}(\pi)$ modulo $\mathcal{N}_L$. This is the content of the following theorem:

(8.1) Theorem. *Let $L|K$ be an abelian extension of degree n and $\mathfrak{p}$ an unramified prime ideal of K.*

If $\pi \in K_{\mathfrak{p}}$ is a prime element, $\overline{\mathfrak{n}}_{\mathfrak{p}}(\pi) \in C_K$ is the idèle class represented by the idèle $\mathfrak{n}_{\mathfrak{p}}(\pi) = (\dots, 1, 1, 1, \pi, 1, 1, 1, \dots)$, and f is the smallest number such that
$$\overline{\mathfrak{n}}_{\mathfrak{p}}(\pi)^f \in N_{L|K} C_L,$$

then the prime ideal $\mathfrak{p}$ factors in the extension L into $r = n/f$ distinct prime ideals $\mathfrak{P}_1, \dots, \mathfrak{P}_r$ of degree f.

Hence if one knows the norm group $N_{L|K} C_L$, then one can simple read off from the idèle class group C_K of the base field K how $\mathfrak{p}$ decomposes in L.

Proof. Since the prime ideal $\mathfrak{p}$ is unramified, it factors in L into distinct prime ideals of equal degree: $\mathfrak{p} = \mathfrak{P}_1 \cdots \mathfrak{P}_r$. By the reciprocity law $C_K / N_{L|K} C_L \cong G_{L|K}$, and $\overline{\mathfrak{n}}_{\mathfrak{p}}(\pi) \bmod N_{L|K} C_L$ in $C_K / N_{L|K} C_L$ has the same order f as (cf. (6.15))
$$(\overline{\mathfrak{n}}_{\mathfrak{p}}(\pi), L|K) = (\pi, L_{\mathfrak{P}}|K_{\mathfrak{p}}) \in G_{L_{\mathfrak{P}}|K_{\mathfrak{p}}} \subseteq G_{L|K} \qquad (\mathfrak{P} \mid \mathfrak{p}).$$
But by II, (4.8) $(\pi, L_{\mathfrak{P}}|K_{\mathfrak{p}}) = \varphi_{\mathfrak{p}}$ is the Frobenius automorphism of the unramified extension $L_{\mathfrak{P}}|K_{\mathfrak{p}}$. It generates the group $G_{L_{\mathfrak{P}}|K_{\mathfrak{p}}}$, thus the order f

---

[24] The terminology "reciprocity law" also comes from this; this might appear a little strange, since at first glance Artin's Reciprocity Law seems to have nothing in common with Gauss's law.

coincides with the degree $[L_{\mathfrak{P}} : K_{\mathfrak{p}}]$, and therefore with the degree of $\mathfrak{P}$. Therefore the number r of distinct prime ideals $\mathfrak{P}_1, \ldots, \mathfrak{P}_r$ over $\mathfrak{p}$ can be computed from this by the fundamental equation of number theory: $n = r \cdot f$.

Later, when we derive the classical, ideal-theoretic theorems of class field theory from our idèle-theoretic theorems, we will encounter Theorem (8.1) again in another, purely ideal theoretic formulation.

Theorem (8.1) only applies to unramified prime ideals. To decide whether a prime ideal $\mathfrak{p}$ is unramified, the following observations are useful.

If $L|K$ is an abelian extension of algebraic number fields, and if $L_{\mathfrak{P}}|K_{\mathfrak{p}}$ are the associated local extensions, then L is uniquely determined by the norm group $N_{L|K}C_L \subseteq C_K$ and $L_{\mathfrak{P}}$ by the norm group $N_{L_{\mathfrak{P}}|K_{\mathfrak{p}}}L_{\mathfrak{P}}^{\times} \subseteq K_{\mathfrak{p}}^{\times}$. There is a simple relation between these two norm groups. To see this, we use (7.3) and consider the group $K_{\mathfrak{p}}^{\times}$ as embedded in C_K via the homomorphism

$$\overline{n}_{\mathfrak{p}} : K_{\mathfrak{p}}^{\times} \longrightarrow C_K,$$

where we identify $x_{\mathfrak{p}} \in K_{\mathfrak{p}}^{\times}$ with the idèle class represented by the idèle $n_{\mathfrak{p}}(x_{\mathfrak{p}}) = (\ldots, 1, 1, 1, x_{\mathfrak{p}}, 1, 1, 1, \ldots)$. Using the abbreviations $N = N_{L|K}$ and $N_{\mathfrak{P}} = N_{L_{\mathfrak{P}}|K_{\mathfrak{p}}}$ we then obtain the

(8.2) Proposition. *If $L|K$ is an abelian extension, then*

$$NC_L \cap K_{\mathfrak{p}}^{\times} = N_{\mathfrak{P}}L_{\mathfrak{P}}^{\times}.$$

Proof. If $x_{\mathfrak{p}} \in N_{\mathfrak{P}}L_{\mathfrak{P}}^{\times}$, then the idèle $n_{\mathfrak{p}}(x_{\mathfrak{p}}) = (\ldots, 1, 1, 1, x_{\mathfrak{p}}, 1, 1, 1, \ldots)$ has only norm components, and is therefore a norm idèle of L by (3.4). Thus $N_{\mathfrak{P}}L_{\mathfrak{P}}^{\times} \subseteq NC_L \cap K_{\mathfrak{p}}^{\times}$.

Conversely, let $\overline{a} \in NC_L \cap K_{\mathfrak{p}}^{\times}$. Then $\overline{a}$ is represented on one hand by the norm idèle $a = N\mathfrak{b}$, $\mathfrak{b} \in I_L$, and on the other hand by an idèle $n_{\mathfrak{p}}(x_{\mathfrak{p}}) = (\ldots, 1, 1, 1, x_{\mathfrak{p}}, 1, 1, 1, \ldots)$, $x_{\mathfrak{p}} \in K_{\mathfrak{p}}^{\times}$, so that

$$n_{\mathfrak{p}}(x_{\mathfrak{p}}) \cdot a = N\mathfrak{b} \qquad \text{with } a \in K^{\times}.$$

Passing to the components, we see that a is a norm for all $\mathfrak{q} \neq \mathfrak{p}$. Therefore it follows from the product formula (6.15) that a is also a norm at the prime $\mathfrak{p}$, so that $x_{\mathfrak{p}} \in N_{\mathfrak{P}}L_{\mathfrak{P}}^{\times}$, which proves the other inclusion $NC_L \cap K^{\times} \subseteq N_{\mathfrak{P}}L_{\mathfrak{P}}^{\times}$.

The following theorem now shows how the ramification of a prime $\mathfrak{p}$ of K in an abelian extension $L|K$ is reflected in the norm group $NC_L \subseteq C_K$. We call an infinite prime $\mathfrak{p}$ unramified if it splits completely, i.e., if $L_{\mathfrak{P}} = K_{\mathfrak{p}}$.

(8.3) Theorem. *Let $L|K$ be an abelian extension, $\mathcal{N} = NC_L \subseteq C_K$ its norm group and $\mathfrak{p}$ a prime of K. Then we have the following equivalences:*

$$\mathfrak{p} \quad \text{is unramified in } L \quad \Longleftrightarrow \quad U_{\mathfrak{p}} \subseteq \mathcal{N},$$
$$\mathfrak{p} \quad \text{splits completely in } L \quad \Longleftrightarrow \quad K_{\mathfrak{p}}^{\times} \subseteq \mathcal{N}.$$

Proof. The prime $\mathfrak{p}$ is unramified $\Leftrightarrow L_{\mathfrak{P}}|K_{\mathfrak{p}}$ is unramified (i.e., $L_{\mathfrak{P}} = K_{\mathfrak{p}}$ for $\mathfrak{p}|\infty$) $\Leftrightarrow$ (by II, (4.9) and (8.2)) $U_{\mathfrak{p}} \subseteq N_{\mathfrak{P}} L_{\mathfrak{P}}^{\times} = \mathcal{N} \cap K_{\mathfrak{p}}^{\times} \Leftrightarrow U_{\mathfrak{p}} \subseteq \mathcal{N}$. Similarly, $\mathfrak{p}$ splits completely $\Leftrightarrow L_{\mathfrak{P}} = K_{\mathfrak{p}} \Leftrightarrow K_{\mathfrak{p}}^{\times} = N_{\mathfrak{P}} L_{\mathfrak{P}}^{\times} = \mathcal{N} \cap K_{\mathfrak{p}}^{\times} \Leftrightarrow K_{\mathfrak{p}}^{\times} \subseteq \mathcal{N}$.

As in the local case, we have in global class field theory the notion of a **conductor**. For a local abelian extension $L_{\mathfrak{P}}|K_{\mathfrak{p}}$ the conductor $\mathfrak{f}_{\mathfrak{p}}$ was defined as the smallest $\mathfrak{p}$-power $\mathfrak{p}^n$ such that $U_{\mathfrak{p}}^n \subseteq N_{\mathfrak{P}} L_{\mathfrak{P}}^{\times}$. In the global case we have to replace the $\mathfrak{p}$-powers $\mathfrak{p}^n$ by the moduli $\mathfrak{m}$ (cf. §7, p.163) and the groups $U_{\mathfrak{p}}^n = \{x_{\mathfrak{p}} \in K_{\mathfrak{p}}^{\times} \mid x_{\mathfrak{p}} \equiv 1 \bmod \mathfrak{p}^n\}$ by the congruence subgroups $C_K^{\mathfrak{m}} = \{\bar{\mathfrak{a}} \in C_K \mid \mathfrak{a} \equiv 1 \bmod \mathfrak{m}\}$.

If we keep in mind that by (7.9) every norm group $\mathcal{N} \subseteq C_K$ contains a congruence subgroup $C_K^{\mathfrak{m}}$, and that $\mathfrak{m} \mid \mathfrak{m}'$ implies the inclusion $C_K^{\mathfrak{m}'} \subseteq C_K^{\mathfrak{m}}$ (and not conversely!), then we come to the following definition.

(8.4) Definition. *Let $L|K$ be an abelian extension with norm group $\mathcal{N} = NC_L$. By the **conductor** $\mathfrak{f}$ of $\mathcal{N}$, or also of $L|K$, we mean the g.c.d. of all moduli $\mathfrak{m}$ such that $C_K^{\mathfrak{m}} \subseteq \mathcal{N}$.*

Thus $C_K^{\mathfrak{f}}$ is the largest congruence subgroup contained in $\mathcal{N}$, and $C_K^{\mathfrak{m}} \subseteq \mathcal{N}$ if and only if $\mathfrak{f} \mid \mathfrak{m}$. Note that $\mathcal{N} = C_K^{\mathfrak{m}}$, i.e. $\mathcal{N}$ is the norm group of the ray class field mod $\mathfrak{m}$, does not imply in general that $\mathfrak{m}$ is the conductor of $\mathcal{N}$ (for example, $C_{\mathbb{Q}}^1 = C_{\mathbb{Q}}^{p\infty}$ by (7.10)).

For the conductor $\mathfrak{f}$ of an abelian extension $L|K$ one has a localization theorem analogous to that for the discriminant. If we define for an infinite prime the conductor $\mathfrak{f}_{\mathfrak{p}}$ by $\mathfrak{p}$ or 1, depending on whether $L_{\mathfrak{P}} \neq K_{\mathfrak{p}}$ or $L_{\mathfrak{P}} = K_{\mathfrak{p}}$, we have

(8.5) Theorem. *If $\mathfrak{f}$ is the conductor of an abelian extension $L|K$ and $\mathfrak{f}_{\mathfrak{p}}$ is the conductor of the local extension $L_{\mathfrak{P}}|K_{\mathfrak{p}}$, then*

$$\mathfrak{f} = \prod_{\mathfrak{p}} \mathfrak{f}_{\mathfrak{p}}.$$

Proof. We have to show: If $\mathcal{N} = NC_L$ and $\mathfrak{m} = \prod_{\mathfrak{p}} \mathfrak{p}^{n_{\mathfrak{p}}}$ is a modulus of K,

$$C_K^{\mathfrak{m}} \subseteq \mathcal{N} \quad \Longleftrightarrow \quad \prod_{\mathfrak{p}} \mathfrak{f}_{\mathfrak{p}} \mid \mathfrak{m} \quad \Longleftrightarrow \quad \mathfrak{f}_{\mathfrak{p}} \mid \mathfrak{p}^{n_{\mathfrak{p}}} \text{ for all } \mathfrak{p}.$$

Using the notation from §7, p. 163 and (8.2), this follows from the equivalences

$$C_K^{\mathfrak{m}} \subseteq \mathcal{N} \iff (\mathfrak{a} \equiv 1 \bmod \mathfrak{m} \implies \bar{\mathfrak{a}} \in \mathcal{N}) \text{ for } \mathfrak{a} \in I_K$$
$$\iff (\mathfrak{a}_{\mathfrak{p}} \equiv 1 \bmod \mathfrak{p}^{n_{\mathfrak{p}}} \implies \bar{\mathfrak{n}}_{\mathfrak{p}}(\mathfrak{a}_{\mathfrak{p}}) \in \mathcal{N} \cap K_{\mathfrak{p}}^{\times} = N_{\mathfrak{P}} L_{\mathfrak{P}}^{\times})$$
$$\iff (\mathfrak{a}_{\mathfrak{p}} \in U_{\mathfrak{p}}^{n_{\mathfrak{p}}} \implies \mathfrak{a}_{\mathfrak{p}} \in N_{\mathfrak{P}} L_{\mathfrak{P}}^{\times})$$
$$\iff U_{\mathfrak{p}}^{n_{\mathfrak{p}}} \subseteq N_{\mathfrak{P}} L_{\mathfrak{P}}^{\times} \quad .$$
$$\iff \mathfrak{f}_{\mathfrak{p}} \mid \mathfrak{p}^{n_{\mathfrak{p}}}.$$

We call an infinite prime $\mathfrak{p}$ of K ramified in L if $L_{\mathfrak{P}} \neq K_{\mathfrak{p}}$. With this definition, we obtain from II, (7.21) the following result.

(8.6) Theorem. *A prime $\mathfrak{p}$ of K is ramified in L if and only if it appears in the conductor $\mathfrak{f}$ of $L|K$.*

If we further call $L|K$ unramified if all finite as well as all infinite primes are unramified, this implies

(8.7) Corollary. *An abelian extension $L|K$ is unramified if and only if its conductor $\mathfrak{f} = 1$.*

In §7, p. 166 we called the ray class field mod 1, i.e., the class field associated with the norm group C_K^1, the **Hilbert class field** over K. We can characterize this field as follows:

(8.8) Theorem. *The Hilbert class field over K is the maximal unramified extension of K.*

Because of the isomorphism $C_K/C_K^1 \cong J_K/P_K$, the degree of the Hilbert class field over K is the ideal class number $h = (J_K : P_K)$ of K (cf. §7, p. 166). Therefore if $h = 1$, which occurs, for example, if $K = \mathbb{Q}$, then every abelian extension of K is ramified, and the Hilbert class field coincides with K.

The following famous theorem has been conjectured by Hilbert but remained unproved for a long time.

(8.9) Principal Ideal Theorem. *In the Hilbert class field over K every ideal $\mathfrak{a}$ of K becomes a principal ideal.*

E. ARTIN used the reciprocity law to reduce the proof of this theorem to a purely group theoretic problem whose solution was given shortly thereafter by PH. FURTWÄNGLER. In the following we explain Artin's reduction.

If we start with K and take its Hilbert's class field K_1, then the Hilbert class field K_2 of K_1, and we continue in this way, we obtain a chain of class fields

$$K = K_0 \subseteq K_1 \subseteq K_2 \subseteq \ldots,$$

the so-called **class field tower**. For this tower of fields we show first:

(8.10) Proposition. *The i-th class field K_i is normal over K, and K_1 is the largest abelian subfield of K_2; in other words:*

The Galois group $G_{K_2|K_1}$ is the commutator subgroup of $G_{K_2|K}$.

Proof. We assume as induction hypothesis K_i, $i \geq 1$, is normal. Let σ be an isomorphism of $K_{i+1}|K$. Then $\sigma K_i = K_i$, and since $K_{i+1}|K_i$ abelian and unramified, the same holds for $\sigma K_{i+1}|\sigma K_i$, i.e., $\sigma K_{i+1}|K_i$. Then by (8.8) $\sigma K_{i+1} \subseteq K_{i+1}$, thus $K_{i+1}|K$ is normal.

Let K' be the maximal abelian extension of K contained in K_2. Then we have $K_1 \subseteq K'$, and $K' \subseteq K_1$, since $K'|K$ is unramified; thus in fact $K' = K_1$.

For the proof of the Principal Ideal Theorem we now have to translate the statement that every ideal of K becomes a principal ideal in the class field K_1 into the language of idèles. This is obviously equivalent to the canonical map

$$J_K/P_K \longrightarrow J_{K_1}/P_{K_1},$$

that takes the ideal $\mathfrak{a} \in J_K$ to the ideal in the field K_1 generated by $\mathfrak{a}$ being trivial. On the other hand, by §7, p. 164 we have the canonical isomorphisms

$$C_K/C_K^1 \cong J_K/P_K \quad \text{and} \quad C_{K_1}/C_{K_1}^1 \cong J_{K_1}/P_{K_1},$$

and therefore the commutative diagram

$$
\begin{array}{ccc}
C_K/C_K^1 & \longrightarrow & J_K/P_K \\
\downarrow{\scriptstyle i} & & \downarrow \\
C_{K_1}/C_{K_1}^1 & \longrightarrow & J_{K_1}/P_{K_1},
\end{array}
$$

where the homomorphism i is induced from the canonical embedding $C_K \to C_{K_1}$. Thus the Principal Ideal Theorem says precisely that the map i is trivial, which is equivalent to the statement that we have an inclusion $C_K \subseteq C_{K_1}^1$.

Since $C_{K_1}^1$ is the norm group of the extension $K_2|K_1$, this means that we simply have to show, making use of the norm residue symbol, that

$$1 = (C_K, K_2|K_1) = \text{Ver}(C_K, K_2|K).$$

Now if we take into account that

$$(C_K, K_2|K) = G^{\mathrm{ab}}_{K_2|K} = G_{K_2|K}/G_{K_2|K_1} = G_{K_1|K},$$

and that $G_{K_2|K_1}$ is the commutator subgroup of $G_{K_2|K}$, then we see that the Principal Ideal Theorem reduces to the following theorem:

(8.11) Theorem. *If G is a metabelian finite group, i.e., a finite group with abelian commutator subgroup G', then the transfer of G to G'* [25)]

$$\mathrm{Ver} : G^{\mathrm{ab}} \longrightarrow G'^{\mathrm{ab}} = G'$$

is the trivial map.

This theorem is purely group theoretic but by no means of a simple nature. It was first proved (1930) by PH. FURTWÄNGLER, a simpler proof was given by S. IYANAGA (1934). It would exceed the scope of these lectures to give this proof, instead we refer to [23].

A problem posed by FURTWÄNGLER, which is closely related to the Principal Ideal Theorem, is the famous **class field tower problem**. It is the question of whether the class field tower

$$K = K_0 \subseteq K_1 \subseteq K_2 \subseteq \dots$$

(K_{i+1} the Hilbert class field over K_i) terminates after finitely many steps. A positive answer to this question would have the following interesting consequence: If $K_{i+1} = K_i$ for sufficiently large i, then K_i has class number 1. One would therefore obtain for every algebraic number field K a canonically given solvable extension, in which not only the ideals of K but all the ideals become principal ideals. This problem was open for a long time until in the year 1964 is was decided in the negative by E.S. GOLOD and I.R. ŠAFAREVIČ who showed that there are in fact infinite class field towers. It is interesting to note that similar to the proof of the Principal Ideal Theorem, this was done using a reduction to a purely group theoretic conjecture which was solved shortly afterwards. We refer the interested reader to [11], IX, and [14].

§ 9. The Ideal Theoretic Formulation of Class Field Theory

The results of global class field theory obtained so far are almost exclusively formulated in terms of idèles. We have seen that the idèle-theoretic language

---

[25)] We recall that the transfer or Verlagerung of a group G to a subgroup g is defined as the restriction $G^{\mathrm{ab}} \cong H^{-2}(G, \mathbb{Z}) \xrightarrow{\mathrm{res}} H^{-2}(g, \mathbb{Z}) \cong g^{\mathrm{ab}}$.

has remarkable technical advantages which justify its central role in our discussions. Now that we have reached some form of conclusion, we should also derive from the results obtained so far the classical purely ideal-theoretic theorems of class field theory as stated, for example, in Hasse's Zahlbericht [17].

In the idèle-theoretic formulation of the reciprocity law, the abelian extensions $L|K$ correspond uniquely to the norm groups $N_{L|K}C_L \subseteq C_K$. In the ideal-theoretic form there is a similar correspondence, however, not quite as simple. Here the abelian extensions also correspond to certain norm groups in the ideal group J_K of K. The norm residue symbol $(\ , L|K) : C_K \to G_{L|K}$ is replaced by a symbol which maps the ideals $\mathfrak{a} \in J_K$ to elements of the Galois group $G_{L|K}$. It is a characteristic of the ideal-theoretic version that such a symbol cannot be defined for all ideals, in contrast to the norm residue symbol, which is defined for all idèles without any restriction. More precisely, the ideals that have to be omitted are the ones which are divisible by ramified primes. This is done by choosing a (sufficiently large) so-called "modulus of definition $\mathfrak{m}$", in which all the ramified prime ideals appear, and to which the ideals being considered are to be relatively prime. Field theoretically this choice corresponds to the embedding of the abelian extension $L|K$ into the ray class field mod $\mathfrak{m}$, in which by §8 only prime divisors of $\mathfrak{m}$ are ramified. This process, namely the choice of a sufficiently large modulus of definition $\mathfrak{m}$ for every given abelian extension $L|K$, the transition to the ideals relatively prime to $\mathfrak{m}$, and the embedding of $L|K$ into the ray class field mod $\mathfrak{m}$, are the crucial inputs to the ideal-theoretic formulation of the reciprocity law. We describe this more precisely below.

We take an algebraic number field K as our base field. By J_K (resp. P_K), we denote again the ideal (resp. principal ideal) group of K; however, since the base field K is always fixed, we omit the index K, just as for the idèle group I_K and the idèle class group C_K, and we write J, P, I, C etc.

If $\mathfrak{m} = \prod_{\mathfrak{p}} \mathfrak{p}^{n_{\mathfrak{p}}}$ is a **modulus** of K, then we let

    $J^{\mathfrak{m}}$ be the group of all ideals relatively prime to $\mathfrak{m}$,

    $P^{\mathfrak{m}}$ the group of all principal ideals $(a) \in P$ with $a \equiv 1 \bmod \mathfrak{m}$ [26].

$P^{\mathfrak{m}}$ is called the **ray mod $\mathfrak{m}$**, and every group between $P^{\mathfrak{m}}$ and $J^{\mathfrak{m}}$ is in the classical terminology called an **ideal group defined mod $\mathfrak{m}$**. For the ideal groups defined mod $\mathfrak{m}$ we use the notation $H^{\mathfrak{m}}$.

The factor group $J^{\mathfrak{m}}/P^{\mathfrak{m}}$ is called the **ray class group mod $\mathfrak{m}$**. Obviously, if $\mathfrak{m} = 1$, then $J^{\mathfrak{m}} = J$ and $P^{\mathfrak{m}} = P$, i.e., we obtain the full ideal class group J/P as the ray class group mod 1.

In place of the idèle class group C we now have a whole family of ray class groups $J^{\mathfrak{m}}/P^{\mathfrak{m}}$. Whereas the idèle class group C was the source group for all

---

[26] This congruence is again to be understood in the sense of §7, p. 164.

abelian extensions of the base field K, the ray class groups $J^{\mathfrak{m}}/P^{\mathfrak{m}}$ are only responsible for the extensions contained in the ray class field mod $\mathfrak{m}$.

If $L|K$ is any extension, then we call a modulus $\mathfrak{m}$ a **modulus of definition for $L\,|\,K$** if L lies in the ray class field mod $\mathfrak{m}$ (i.e., $C^{\mathfrak{m}} \subseteq N_{L|K}C_L$). If $\mathfrak{m} \mid \mathfrak{m}'$ and if $\mathfrak{m}$ is a modulus of definition for $L|K$, then so is $\mathfrak{m}'$; by definition, the conductor of $L|K$ is the g.c.d. of all the moduli of definition (cf. (8.4)).

After choosing a modulus of definition $\mathfrak{m}$, we now associate with every abelian extension $L|K$ the following ideal group defined mod $\mathfrak{m}$:

(9.1) Definition. *If $L|K$ is an abelian extension and $\mathfrak{m}$ is a modulus of definition for $L|K$, then*
$$H^{\mathfrak{m}} = N_{L|K}J_L^{\mathfrak{m}} \cdot P^{\mathfrak{m}}$$
*is called the **ideal group defined mod $\mathfrak{m}$** associated with $L|K$. Here $J_L^{\mathfrak{m}}$ denotes the group of all ideals of L relatively prime to $\mathfrak{m}$.*

The map from the norm group $H^{\mathfrak{m}}/P^{\mathfrak{m}}$ into the ray class group $J^{\mathfrak{m}}/P^{\mathfrak{m}}$ is the ideal-theoretic analogue of the map from the norm group $N_{L|K}C_L$ into the idèle class group C_K.

In place of the norm residue symbol we now define a homomorphism
$$J^{\mathfrak{m}} \xrightarrow{\left(\frac{L|K}{\quad}\right)} G_{L|K},$$
by mapping each ideal $\mathfrak{a}$ of K relatively prime to $\mathfrak{m}$ to an automorphism $\left(\frac{L|K}{\mathfrak{a}}\right)$ in $G_{L|K}$, which is called the **Artin symbol**. By multiplicativity, it suffices to consider the prime ideals $\mathfrak{p}$ of K which do not divide $\mathfrak{m}$. For these we set
$$\left(\frac{L|K}{\mathfrak{p}}\right) = \varphi_{\mathfrak{p}} \in G_{L|K},$$
where $\varphi_{\mathfrak{p}}$ is the **Frobenius automorphism** of $L|K$ associated with $\mathfrak{p}$. We briefly recall its definition: If $\mathfrak{P}$ is a prime of L lying over $\mathfrak{p}$, then $\varphi_{\mathfrak{p}}$ is an element of the decomposition group $G_{\mathfrak{P}} = G_{L_{\mathfrak{P}}|K_{\mathfrak{p}}} \subseteq G_{L|K}$ of $\mathfrak{P}$ over K and is as such (because there is no ramification) uniquely determined by
$$\varphi_{\mathfrak{p}}a \equiv a^q \mod \mathfrak{P} \quad \text{for all integers } a \in L.$$
Here q is the number of elements of the residue field of $\mathfrak{p}$; $\varphi_{\mathfrak{p}}$ does not depend on the choice of the prime ideals $\mathfrak{P}$ but only on $\mathfrak{p}$, since $G_{L|K}$ is abelian and an ideal conjugate to $\mathfrak{P}$ yields a Frobenius automorphism conjugate to $\varphi_{\mathfrak{p}}$.

The **Artin Reciprocity Law** in its classical ideal-theoretic form now reads as follows:

(9.2) Theorem. *If $L|K$ is an abelian extension and $\mathfrak{m}$ is a modulus of definition for $L|K$, then the Artin symbol yields the exact sequence*

$$1 \longrightarrow H^{\mathfrak{m}}/P^{\mathfrak{m}} \longrightarrow J^{\mathfrak{m}}/P^{\mathfrak{m}} \xrightarrow{\left(\frac{L|K}{\cdot}\right)} G_{L|K} \longrightarrow 1,$$

where $J^{\mathfrak{m}}$ is the group of all ideals of K relatively prime to $\mathfrak{m}$ and $H^{\mathfrak{m}}$ is the ideal group defined mod $\mathfrak{m}$ associated with L (cf. (9.1)).

Remark. Note that this implies in particular that the Artin symbol does not depend on the ideals themselves but only on the ideal classes mod $P^{\mathfrak{m}}$; therefore it induces a homomorphism on the class group $J^{\mathfrak{m}}/P^{\mathfrak{m}} \longrightarrow G_{L|K}$.

We will prove the exactness of the sequence

$$1 \longrightarrow H^{\mathfrak{m}}/P^{\mathfrak{m}} \longrightarrow J^{\mathfrak{m}}/P^{\mathfrak{m}} \xrightarrow{\left(\frac{L|K}{\cdot}\right)} G_{L|K} \longrightarrow 1$$

by comparing it with the analogous idèle-theoretic sequence

$$1 \longrightarrow N_{L|K}C_L \longrightarrow C_K \xrightarrow{(\ ,L|K)} G_{L|K} \longrightarrow 1$$

which is exact. More precisely, we will compare it with the exact sequence

$$1 \longrightarrow N_{L|K}C_L/C^{\mathfrak{m}} \longrightarrow C/C^{\mathfrak{m}} \xrightarrow{(\ ,L|K)} G_{L|K} \longrightarrow 1,$$

where we pass from idèles to ideals as in (2.3), using the homomorphism

$$\kappa : I \longrightarrow J, \quad \mathfrak{a} \longmapsto \prod_{\mathfrak{p} \nmid \infty} \mathfrak{p}^{v_{\mathfrak{p}}(\mathfrak{a}_{\mathfrak{p}})}.$$

For this homomorphism κ we now have the

(9.3) Proposition. *The map κ induces a canonical isomorphism of ray class groups*
$$\overline{\kappa}_{\mathfrak{m}} : C/C^{\mathfrak{m}} \longrightarrow J^{\mathfrak{m}}/P^{\mathfrak{m}},$$

whose restriction to $N_{L|K}C_L/C^{\mathfrak{m}}$ yields an isomorphism

$$\overline{\kappa}_{\mathfrak{m}} : N_{L|K}C_L/C^{\mathfrak{m}} \longrightarrow H^{\mathfrak{m}}/P^{\mathfrak{m}}.$$

Proof. To prove $\overline{\kappa}_{\mathfrak{m}}$ is an isomorphism, we start with the isomorphism
$$C/C^{\mathfrak{m}} \cong I/I^{\mathfrak{m}} \cdot K^{\times}.$$

Thus we need to find for every idèle class $\mathfrak{a} \cdot I^{\mathfrak{m}} \cdot K^{\times}$ a representative idèle $\mathfrak{a}$, which is mapped under κ to the group $J^{\mathfrak{m}}$ of ideals relatively prime to $\mathfrak{m}$. We find these representatives in the group

$$I^{\langle \mathfrak{m} \rangle} = \{\mathfrak{a} \in I \mid \mathfrak{a}_{\mathfrak{p}} = 1 \text{ for all } \mathfrak{p} \mid \mathfrak{m}\}.$$

In fact, if $\mathfrak{a} \in I$, then by the Approximation Theorem there is an element $a \in K^\times$ such that $\mathfrak{a}_\mathfrak{p} \cdot a \equiv 1 \bmod \mathfrak{p}^{n_\mathfrak{p}}$ for all $\mathfrak{p} \mid \mathfrak{m} = \prod_\mathfrak{p} \mathfrak{p}^{n_\mathfrak{p}}$. Therefore we can write $\mathfrak{a} \cdot a = \mathfrak{a}' \cdot \mathfrak{b}$, where $\mathfrak{a}'$, resp. $\mathfrak{b}$, is defined by the components

$$\mathfrak{a}'_\mathfrak{p} = 1 \text{ for } \mathfrak{p} \mid \mathfrak{m}, \quad \mathfrak{a}'_\mathfrak{p} = \mathfrak{a}_\mathfrak{p} \cdot a \text{ for } \mathfrak{p} \nmid \mathfrak{m}, \text{ resp.}$$

$$\mathfrak{b}_\mathfrak{p} = \mathfrak{a}_\mathfrak{p} \cdot a \equiv 1 \bmod \mathfrak{p}^{n_\mathfrak{p}} \text{ for } \mathfrak{p} \mid \mathfrak{m}, \quad \mathfrak{b}_\mathfrak{p} = 1 \text{ for } \mathfrak{p} \nmid \mathfrak{m}.$$

Since $\mathfrak{a}' \in I^{\langle \mathfrak{m} \rangle}$ and $\mathfrak{b} \in I^\mathfrak{m}$, we have $\mathfrak{a} \in I^{\langle \mathfrak{m} \rangle} \cdot I^\mathfrak{m} \cdot K^\times$. The isomorphism $\overline{\kappa}_\mathfrak{m}$ results now from the fact that the homomorphism

$$I^{\langle \mathfrak{m} \rangle} \longrightarrow J^\mathfrak{m}/P^\mathfrak{m}, \quad \mathfrak{a} \longmapsto \kappa(\mathfrak{a}) \cdot P^\mathfrak{m}$$

is surjective and has kernel $I^\mathfrak{m} \cdot K^\times \cap I^{\langle \mathfrak{m} \rangle}$, which is easy to verify. Hence

$$C/C^\mathfrak{m} \cong I^{\langle \mathfrak{m} \rangle} \cdot I^\mathfrak{m} \cdot K^\times / I^\mathfrak{m} \cdot K^\times \cong I^{\langle \mathfrak{m} \rangle}/I^\mathfrak{m} \cdot K^\times \cap I^{\langle \mathfrak{m} \rangle}.$$

For the second claim we have to show further that

$$\overline{\kappa}_\mathfrak{m}(N_{L|K} C_L / C^\mathfrak{m}) = N_{L|K} J_L^\mathfrak{m} \cdot P^\mathfrak{m}/P^\mathfrak{m}.$$

If we set $I_L^{\langle \mathfrak{m} \rangle} = \{ \mathfrak{a} \in I_L \mid \mathfrak{a}_\mathfrak{P} = 1 \text{ for } \mathfrak{P} \mid \mathfrak{m} \} \subseteq I_L$, then we have from the above $I_L = I_L^{\langle \mathfrak{m} \rangle} \cdot I_L^\mathfrak{m} \cdot L^\times$, and therefore

$$N_{L|K} C_L / C^\mathfrak{m} = N_{L|K} I_L \cdot K^\times / I^\mathfrak{m} \cdot K^\times = N_{L|K} I_L^{\langle \mathfrak{m} \rangle} \cdot I^\mathfrak{m} \cdot K^\times / I^\mathfrak{m} \cdot K^\times.$$

From the definition of $\overline{\kappa}_\mathfrak{m}$ it follows that

$$\overline{\kappa}_\mathfrak{m}(N_{L|K} C_L / C^\mathfrak{m}) = \kappa(N_{L|K} I_L^{\langle \mathfrak{m} \rangle}) \cdot P^\mathfrak{m}/P^\mathfrak{m} = N_{L|K}(\kappa(I_L^{\langle \mathfrak{m} \rangle})) \cdot P^\mathfrak{m}/P^\mathfrak{m},$$

and because $\kappa(I_L^{\langle \mathfrak{m} \rangle}) = J_L^\mathfrak{m}$, we obtain the desired result.

The isomorphism $\overline{\kappa}_\mathfrak{m} : C/C^\mathfrak{m} \longrightarrow J^\mathfrak{m}/P^\mathfrak{m}$ yields a surjective homomorphism

$$\kappa_\mathfrak{m} : C \longrightarrow J^\mathfrak{m}/P^\mathfrak{m}$$

with kernel $C^\mathfrak{m}$. Since $J^\mathfrak{m}$ is generated by the prime ideals $\mathfrak{p} \nmid \mathfrak{m}$, we can describe the map $\kappa_\mathfrak{m}$ as follows.

Let $\mathfrak{p} \nmid \mathfrak{m}$ be a prime ideal of K, $\pi \in K_\mathfrak{p}$ a prime element in $K_\mathfrak{p}$, and $\mathfrak{n}_\mathfrak{p}(\pi) = (\dots, 1, 1, \pi, 1, 1, \dots)$ the "prime idèle" associated with $\mathfrak{p}$ in the idèle class $\overline{\mathfrak{n}}_\mathfrak{p}(\pi) = \mathfrak{n}_\mathfrak{p}(\pi) \cdot K^\times \in C$. Then the map $\kappa_\mathfrak{m}$ takes the idèle class $\overline{\mathfrak{n}}_\mathfrak{p}(\pi)$ to the class $\mathfrak{p} \cdot P^\mathfrak{m}$, since $\kappa(\mathfrak{n}_\mathfrak{p}(\pi)) = \mathfrak{p}$.

Theorem (9.2) is now an immediate consequence of the following theorem which establishes a relation between the idèle-theoretic and the ideal-theoretic reciprocity law.

(9.4) Theorem. *Let $L|K$ be an abelian extension, and let $\mathfrak{m}$ be a modulus of definition for $L|K$. Then the diagram*

$$
\begin{array}{ccccccccc}
1 & \longrightarrow & N_{L|K}C_L & \longrightarrow & C_K & \xrightarrow{(\;,L|K)} & G_{L|K} & \longrightarrow & 1 \\
 & & \kappa_{\mathfrak{m}}\downarrow & & \kappa_{\mathfrak{m}}\downarrow & & \downarrow{\scriptstyle \mathrm{id}} & & \\
1 & \longrightarrow & H^{\mathfrak{m}}/P^{\mathfrak{m}} & \longrightarrow & J^{\mathfrak{m}}/P^{\mathfrak{m}} & \xrightarrow{(\frac{L|K}{\cdot})} & G_{L|K} & \longrightarrow & 1
\end{array}
$$

commutes, and the surjective homomorphisms $\kappa_{\mathfrak{m}}$ both have kernel $C_K^{\mathfrak{m}}$.

Addendum: *If $\mathfrak{p} \nmid \mathfrak{m}$ is a prime ideal of K, $\pi \in K_{\mathfrak{p}}$ is a prime element, and $\mathfrak{n}_{\mathfrak{p}}(\pi) = (\ldots, 1, 1, \pi, 1, 1, \ldots)$ is the prime idèle associated with $\mathfrak{p}$ in the idèle class $\overline{\mathfrak{n}}_{\mathfrak{p}}(\pi) \in C_K$, then the Artin and the norm residue symbols satisfy*

$$
\left(\frac{L|K}{\mathfrak{p}}\right) = \left(\overline{\mathfrak{n}}_{\mathfrak{p}}(\pi), L|K\right).
$$

Proof.   The equality follows easily from the product formula for the norm residue symbol (6.15) and II, (4.8):

$$
\left(\frac{L|K}{\mathfrak{p}}\right) = \varphi_{\mathfrak{p}} = (\pi, L_{\mathfrak{P}}|K_{\mathfrak{p}}) = \left(\overline{\mathfrak{n}}_{\mathfrak{p}}(\pi), L|K\right).
$$

Because $\kappa_{\mathfrak{m}}(\overline{\mathfrak{n}}_{\mathfrak{p}}(\pi)) = \mathfrak{p}$, it follows that the diagram commutes.

In this section we always considered only unramified prime ideals and excluded the ramified primes by choosing a modulus of definition $\mathfrak{m}$, thus embedding the field in question into a suitable ray class field. It is easy to see that this restriction is necessary. The Artin symbol $(\frac{L|K}{\mathfrak{p}})$, which for an unramified prime ideal $\mathfrak{p}$ is defined by the above Addendum as the norm residue symbol $(\overline{\mathfrak{n}}_{\mathfrak{p}}(\pi), L|K) = (\pi, L_{\mathfrak{P}}|K_{\mathfrak{p}})$, does not permit such a definition in the ramified case, because the norm residue symbol still depends on the choice of the prime element π. The inclusion of the ramified primes into class field theoretical required the step from ideals to idèles, which allows to reduce the global to the local theory, where one can deal with ramified extensions using cohomological methods.

To end, we want to formulate the decomposition law for the unramified prime ideals in an abelian extension $L|K$ in terms of the corresponding ideal group defined mod $\mathfrak{m}$ $H^{\mathfrak{m}}$, which determines the field L as a class field. This gives the ideal-theoretic formulation of Theorem (8.1).

(9.5) Theorem. *Let $L|K$ be an abelian extension and $\mathfrak{p}$ an unramified prime ideal of K. Let further $\mathfrak{m}$ be a modulus of definition for $L|K$ not divisible by $\mathfrak{p}$ (say the conductor) and let $H^{\mathfrak{m}}$ be the corresponding ideal group. If f is the order of $\mathfrak{p}$ mod $H^{\mathfrak{m}}$ in the class group $J^{\mathfrak{m}}/H^{\mathfrak{m}}$, and therefore the smallest number such that*

$$\mathfrak{p}^f \in H^{\mathfrak{m}},$$

then $\mathfrak{p}$ factors in the extension field L into exactly $r = [L : K]/f$ distinct prime ideals $\mathfrak{P}_1, \ldots, \mathfrak{P}_r$ of equal degree f over $\mathfrak{p}$.

Proof. Let $\mathfrak{p} = \mathfrak{P}_1 \cdots \mathfrak{P}_r$ is the prime decomposition of $\mathfrak{p}$ in L. Because $\mathfrak{p}$ is unramified, the $\mathfrak{P}_1, \ldots, \mathfrak{P}_r$ are distinct and of equal degree f over $\mathfrak{p}$. This degree coincides with the order of the decomposition group of $\mathfrak{P}_i$ over K, and therefore with the order of the Frobenius automorphism $\varphi_{\mathfrak{p}} \in G_{L|K}$, which generates the decomposition group. Under the isomorphism

$$J^{\mathfrak{m}}/H^{\mathfrak{m}} \cong G_{L|K}$$

induced by the Artin symbol the element $\varphi_{\mathfrak{p}} = \left(\frac{L|K}{\mathfrak{p}} \right)$ corresponds to the class $\mathfrak{p} \cdot H^{\mathfrak{m}} \in J^{\mathfrak{m}}/H^{\mathfrak{m}}$, hence this class has order f.   q.e.d.

References

1. ALBERT, A. Structures of Algebras. A.M.S. Publ., Providence (1961).
2. ARTIN, E. & TATE, J. Class Field Theory. Harvard (1961).
3. ARTIN, E. Die gruppentheoretische Struktur der Diskriminanten algebraischer Zahlkörper. J. reine und angew. Math. **164**, 1 – 11, (1931).
4. ARTIN, E. Kohomologie endlicher Gruppen. Universität Hamburg (1957).
5. ARTIN, E. Theory of Algebraic Numbers. Lecture notes, Göttingen (1959).
6. BOREVIČ, S.I. & ŠAFAREVIČ, I.R. Zahlentheorie. Birkhäuser Verlag, Basel-Stuttgart (1966).
7. BOURBAKI, N. Algèbre. Ch. V. Hermann, Paris (1950).
8. BOURBAKI, N. Algèbre commutative. Hermann, Paris (1964).
9. BOURBAKI, N. Topologie générale, Ch. I, II, III. Hermann, Paris (1961).
10. CARTAN, H. & EILENBERG, S. Homological Algebra. Princeton Math. Ser. N° 19, (1956).
11. CASSELS, J.W.S. & FRÖHLICH, A. Algebraic Number Theory. Thompson Book Comp. Inc. Washington D.C. (1967).
12. CHEVALLEY, C. Class Field Theory. Universität Nagoya (1954).
13. ENDLER, O. Bewertungstheorie. Bonner Math. Schr. **15**, Bd. 1, 2, (1963).
14. GOLOD, E.S. & ŠAFAREVIČ, I.R. Über Klassenkörpertürme (russisch). Izv. Akad. Nauk. SSSR, **28**, 261 – 272, (1964). Engl. Übersetzung in A.M.S. Transl. (2) **48**, 91 – 102.
15. GROTHENDIECK, A. Sur quelques points d'algèbre homologique. Tohoku Math. J. **9**, 119 – 221, (1957).
16. HALL, M. (JR.) The Theory of Groups. MacMillan, New York (1959).
17. HASSE, H. Bericht über neuere Untersuchungen und Probleme aus der Theorie der algebraischen Zahlkörper. Jahresber. der D. Math. Ver. **35**, (1926), **36**, (1927); **39**, (1930).
18. HASSE, H. Die Normresttheorie relativ-abelscher Zahlkörper als Klassenkörpertheorie im Kleinen. J. reine u. angew. Math. **162**, 145 – 154, (1930).
19. HASSE, H. Die Struktur der R. Brauerschen Algebrenklassengruppe über einem algebraischen Zahlkörper. Math. Annalen, **107**, 248 – 252, (1933).
20. HASSE, H. Führer, Diskriminante und Verzweigungskörper abelscher Zahlkörper. J. reine und angew. Math. **162**, 169 – 184, (1930).
21. HASSE, H. Zahlentheorie. Akademie Verlag Berlin (1963).

22. HILBERT, D. Die Theorie der algebraischen Zahlkörper. Jahresber. der D. Math. Ver. **4**, (1897).
23. IYANAGA, S. Zum Beweis des Hauptidealsatzes. Hamb. Abh. **10**, 349 – 357, (1934).
24. KAWADA, Y. Class Formations. Duke math. J. **22** (1955).
25. KAWADA, Y. Class Formations II. (With SATAKE, ICHIRO). J. Fac. Sci. Univ. Tokyo, Sect. I, 7, (1956).
26. KAWADA, Y. Class Formations III, IV. J. Math. Soc. Japan **7**, (1955), **9**, (1957).
27. KRULL, W. Galoissche Theorie der unendlichen algebraischen Erweiterungen. Math. Annalen **100**, 687 – 698, (1928).
28. KRULL, W. Zur Theorie der Gruppen mit Untergruppentopologie. Hamb. Abh. **28**, 50 – 97, (1965).
29. LANG, S. Algebra. Addison-Wesley, Reading, Mass. (1965).
30. LANG, S. Algebraic Numbers. Addison-Wesley, Reading, Mass. (1964).
31. LANG, S. Rapport sur la Cohomologie des Groupes. Benjamin, New York (1966).
32. LAZARD, M. Sur les groupes de Lie formel à un paramètre. Bull. Soc. Math. France **83**, 251 – 274, (1955).
33. LUBIN, J. One parameter formal Lie groups over p-adic integer rings. Annals of Math. **80**, 464 – 484, (1964).
34. LUBIN, J. & TATE, J. Formal Complex Multiplication in Local Fields. Annals of Math. **81**, 380 – 387, (1965).
35. MACLANE, S. Homology. Springer-Verlag, Berlin-Göttingen-Heidelberg (1963).
36. POITOU, G. Cohomologie Galoisienne des Modules finis. Dunod, Paris (1967).
37. PONTRYAGIN, L. Topologische Gruppen. Teile 1, 2. Teubner, Leipzig (1957–58).
38. SCHILLING, O.F.G. The Theory of Valuations. Math. Surveys IV, New York (1950).
39. SHIMURA, G. & TANIYAMA, Y. Complex Multiplication of Abelian Varieties. Publ. Math. Soc. Japan **6**, (1961).
40. SERRE, J.-P. Abelian ℓ-adic Representations and Elliptic Curves. Benjamin, New York (1968).
41. SERRE, J.-P. Cohomologie Galoisienne. Lecture Notes in Math. **5**, Springer-Verlag, Berlin-Heidelberg-New York (1964).
42. SERRE, J.-P. Corps locaux. Hermann, Paris (1962).
43. SERRE, J.-P. Groupes algebriques et corps de classes. Hermann, Paris (1959).
44. STÖHR, K.-O. Homotopietheorie pro-algebraischer Gruppen und lokale Klassenkörpertheorie. Universität Bonn (1967).
45. TATE, J. The Higher Dimensional Cohomology Groups of Class Field Theory. Ann. of Math. **56**, 294 – 297, (1952).
46. V.D. WAERDEN, B.L. Algebra. Springer-Verlag, 1. Teil (4. Auflage) Berlin (1955), 2. Teil (4. Auflage) Berlin (1959).
47. WEIL, A. Basic Number Theory. Springer-Verlag, New York (1967).
48. WEISS, E. Algebraic Number Theory. McGraw-Hill, New York (1963).
49. ZARISKI, O. & SAMUEL, P. Commutative Algebra, I – II. Van Nostrand, New York (1958 – 60).

Index

abelianization, 20
absolute norm, 79
absolute value, 79, 158
anticommutativity, 25, 49
Artin symbol, 175, 178
augmentation, 4
augmentation ideal, 4

Brauer group, 78

character group, 17, 31
class field, 76, 97, 154, 172
class field tower, 172
class field tower problem, 173
class field, absolute, 166
class field, Hilbert, 166, 171
class formation, 63, 67, 70, 85, 92, 151
class number, 114
coaugmentation, 4
cohomology group, 15
cohomology sequence
   exact, 22
cohomology, trivial, 28
components, local, 125
conductor, 109, 170
congruence subgroup, 164
connecting homomorphism, 21
corestriction, 38, 39
cup product, 44
cyclotomic field, 127, 166

decomposition law, 168, 178
dimension shifting, 30

Existence Theorem, 75, 96, 162

factor systems, 18
field formation, 65, 126, 147
fixed group, 6
formation, 64
Frobenius automorphism, 83, 88, 175
fundamental class, 69, 92, 152

G-free, 9
G-homomorphism, 6
G-induced modules, 27
G-module, 3
G/g-induced, 43
group extensions, 18
group ring, 3

Herbrand module, 53, 129
Herbrand quotient, 53, 129
higher unit group, 79
Hilbert's Theorem 90, 77
homology groups, 15
homomorphisms, crossed, 17

ideal class group, 114
ideal group, 174
idèle class group, 118
idèle class invariant, 141
idèle invariant, 138, 141
idèle topology, 156
idèles, 117
inequality, first fundamental, 129
inequality, second fundamental, 89, 131
inertia field, 85

J. Neukirch, *Class Field Theory*, DOI 10.1007/978-3-642-35437-3,
© Springer-Verlag Berlin Heidelberg 2013